Earth
Rising

T0099480

ECOLOGICAL BELIEF IN AN AGE OF SCIENCE

David Oates

Oregon State University Press
Corvallis, Oregon

The illustration on page 114 is reprinted from *General System Theory* by Ludwig von Bertalanffy with permission of George Braziller, Publishers.
© 1968 by Ludwig von Bertalanffy. All rights reserved.

Photo on back cover of paperback edition by Christiane Covington.

The paper in this book meets the guidelines for permanence and durability of the Committee on Production Guidelines for Book Longevity of the Council on Library Resources and the minimum requirements of the American National Standard for Permanence of Paper for Printed Library Materials z39.48-1984.

The author and publisher would have liked to print this book on paper which was both acid-free and recycled. When this proved impossible, the publisher's ongoing commitment to producing its books on acid-free paper took priority.

Library of Congress Cataloging-in-Publication Data

Oates, David, 1950-

Earth rising: ecological belief in an age of science/David Oates.

p. cm.

Bibliography: p.

Includes index.

ISBN 0-87071-358-2 (cloth), -357-4 (paper) (alk. paper)

1. Ecology—Philosophy. 2. Ecology—Social aspects. I. Title.

QH540.7.027 1989

574.5'01— DC 19

88-25468

CIP

© 1989 David Oates

All rights reserved

Printed in the United States of America

To my parents, with love:
Walter L. Oates
Nancy Belyea Oates

Acknowledgments

The author owes a special debt of gratitude to Jo Alexander, Managing Editor of OSU Press. Her encouragement and commitment played a central role in bringing this book to press. And her informed, spirited editing not only improved the book, but brought a rare warmth to the labors of scholarship.

To my friend Mark Hoist, who helped and encouraged always, a heartfelt thank-you as well.

Contents

"If both empire and state are too large for the exercise of genuine topophilia, it is paradoxical to reflect that the earth itself may eventually command such attachment: this possibility exists because the earth is clearly a natural unit and it has a common history. Shakespeare's words "this blessed plot," "this precious stone set in a silver sea" are not inappropriately applied to the planet itself. Possibly, in some ideal future, our loyalty will be given only to the home region of intimate memories and, at the other end of the scale, to the whole earth."

Yi-Fu Tuan, *Topophilia*

INTRODUCTION
A World to Live In

When, as humans, we find ourselves presented with the varied phenomena of nature—the vast array of stars wheeling overhead, the profusion of animal life, the cyclically dying and returning forests and savannahs, the rocks and streams and hills—it is our instinct to try to find there one meaning that holds it together, that explains it. The same impulse applies to the potentially more confusing inward world of the human, with its dreams, its fears, its experiences of love and awe, its rational and emotional moments, its capacities for logic and story. This reality, too, needs to be included in the single embrace of the world picture. We need to find order in the dancing kaleidoscope of existence, to make the many into One. We need it to make sense.

The classical word for the order and unity of the whole world is *cosmos*. It includes the invisible spirit as well as the tangible earth and skies. Cosmos traditionally means the divinely made world brought out of the primordial chaos, the helter-skelter of disorderly matter. The most familiar symbol for this original formlessness is the restless, ever-changing sea: as in Genesis, the creator spirit "moves upon the face of the deep," and dry land appears—the solid form of reality. Understood this way, the world is organized not just physically but also invisibly, by a principle of order that reaches from the furthest star down to the least impulse of the heart.

Each new generation must be taught how to construct the world, because a cosmos is a human invention, or more precisely an artifice of culture. It takes many tellers, many generations, to fill in the large canvas of existence in a way that both explains the external reality and answers the inner needs of the tellers for order, symmetry, and meaning. A cosmos is not a "given" but an achievement, a distinctive characteristic of our species and its crowning creation.

The differences between one cosmos and another—those elements, say, that distinguish the world of the Middle Ages from the world of the Babylonians—are from this viewpoint less striking than the similarities. People believe in many gods, or in one, or in none; in science or in religion or in politics; but, consciously or unconsciously, all seek some way to understand the world and to feel at home in it. In spite of widely divergent ways of seeing the world, we are united with our most distant kin by our common demand for a comprehensive world-harmony.

In its normal state, humanity quite readily constructs these worldviews, these embracing harmonies. In the hunting cultures and growing cultures which occupy the vast majority of humankind's years on the planet, life has virtually always been organized this way. Always, as Mircea Eliade has so vividly described, the so-called "primitive" cultures arrange the world into a harmony of human and natural and supernatural.[1] All parts of existence are linked into a sacred pattern: the rules and mores of the tribe, the right way to grow or gather food, the identity of the sexes, the plants and animals, the earth and stars, the origins of things. And always the essential mysteries—the strangeness of nature, the inexplicability of life and death, the relation of dream and reality—are tamed by their containment within a sense of meaning, a cosmos.

❍ ❍ ❍

I do not mean to create my own myth of origins here: this long experience of our race is no Eden of innocence. Bloodshed, anxiety, and confusion have surely been the common lot for people in this "normal" state, as they are for us. Yet the modern world has found it increasingly difficult to maintain a sense of the cosmos that contains all the confusion and multiplicity. When the accumulation of wealth and population adds to a certain point, the conviction and certainty which characterize unreflective belief in a traditional cosmos may unravel. Those living at an urban crossroads, for instance, find strange beliefs brought into their midst. The panorama of peoples and ideas inevitably brings a sophistication which questions the primal world-stories, and thereby threatens the very order of the world itself. This is the challenge of all urban civilizations.

We are apt to speak of this as the "modern predicament," as if it were a recent invention. But surely the Preacher's world-weary skepticism in *Ecclestiastes* is just this "modern" experience—the failure of simple faith under the weight of knowledge and experience. Omar Khayyam (and not only in his nineteenth-century mistranslation) and sophisticates in many ages have shared this experience. All have sensed the primal oneness breaking into fragments, leaving the solitary knower, standing in the ugly hubbub of the marketplace perhaps, or out alone under the night sky, pondering the emptiness and vanity of life. In the words of another Old Testament writer, "Where there is no vision the people perish."

The worldview lies at the center of consciousness, an organizing power drawing all elements into stable constellation. And without its wholeness neither individual nor culture can thrive.

The Ecological Worldview

This book is about a modern cosmos—a worldview—that has arisen in answer to the human need for a unifying sense of order. This worldview originated in ecology: the political fad of the late sixties and early seventies, the social movement of a few hippies, the abiding commitment of some activists, the biological specialists' word which became a new tenet of American belief. It is a worldview born out of fearful problems of pollution and overpopulation, and nourished by a popular hope that the "new" science of ecology could solve them.

But the pattern of thinking called ecology has proved to have potentials far beyond the science itself. The ecosystem has been found to be a realm of profound connectedness and relation between each of its living things. Ecologically understood, nothing in nature is brutal or ugly: all is unified, contained within a system which conduces to the good of life. Ecology speaks of a natural world of deeply *humane* characteristics—if this self-congratulatory word may be used here. In this world, cooperation, interconnectedness, even "mutual aid" have been discovered in abundance.

These promising discoveries about nature have been extended to conclusions not only about the physical world but about the intangible world as well. Human relations, ethics and morality, social questions, even aesthetics, philosophy, and theology, all have been recast in ecological form—that is, in ways which claim "nature" for their validation and starting point, and which apply the principles of the ecosystem to human life and thought. The ecological worldview begins with a factual science, but ends with conclusions about values.

○ ○ ○

A worldview is a useful generalization. I define it as the way people who share similar assumptions go about solving similar existential and intellectual problems. ("Solving" means making significant, i.e. harmonizing on a large scale not only factually, but emotionally.) When a number of people use the same logical apparatus, and are working to solve the same problem, the result is a reasonably coherent body of belief.

Any worldview must try to solve the basic threats of evil, suffering, and death, and will define and perceive them in specific ways. For the late nineteenth- and early twentieth-century worldview known as Social Darwinism, for example, the natural facts of struggle and selfishness and bloodshed were the central problems to be solved. Social Darwinism used the logic of evolution as its primary way of thinking about these problems, and hence found "progress" to be an important justification for the way things are.

The worldview of ecology uses the logic of the ecosystem as its primary way of thinking. By relating the parts to the whole, it attempts to find the meaning and beauty of facts that are by themselves difficult, brutal, or ugly. Since the "wholes" to which it refers are regarded as definitive and real, the task of finding embracing harmonies is relatively easy. The pattern of ecological connectedness and cooperation is readily applied to human affairs.

At its core, the ecological worldview relies on a few fundamental concepts derived directly from the scientific study of ecosystems: holism, balance, cooperation, and the cybernetic system. But the ecological worldview is not a science: it is a belief system extrapolated from one. In chapters that follow, the worldview is always kept distinct from the science by naming it "ecologism," "ecological (or ecologistic) thinking," or simply, "the worldview of ecology."

Ecological thinking characteristically originates in a genuine scientific insight or concept. But as a popular worldview, ecologism often does not keep pace with changes and refinements in the scientific picture. The worldview will thus occasionally appall the scientist, who will notice old formulations, minority viewpoints, or oversimplifications. This is typical in the relations between science and culture. Though the scientific base of ecologism is real, it is not necessarily *au courant*. But beyond this cautionary note, it may be worth remembering that the overall message of the ecological worldview is amply supported by ecological science: that the human species must realistically acknowledge its place within the natural order.

This ecological worldview is, I think, more widespread and pervasive than is commonly recognized. Recent attempts to undo environmental regulation have not been more than marginally successful. A regard for wilderness and for threatened species has become a familiar and relatively noncontroversial attitude. And the importance of the environment to the health of the human species has become incontestable.

The ecological awakening has permanently changed the way most people think about the human species and nature. And while such an important change is certainly not made overnight, and not accomplished without much conflict with other beliefs, the evidence is strong that a powerful new influence has emerged in the Western mind: the ecological worldview.

A Few Words about Method

I do not write as a scientific expert, but as a citizen of my culture and era. We live at a fascinating moment, a rare time when the place of humans in the natural world is being questioned and reformulated. New ideas of how to understand the living process are being introduced; and these, unexpectedly, are leading many thinkers around to some very old ideas about the almost sacred importance of the earth and our connection to it. These ideas concern all of us, for they represent not only an attempt to solve immediate ecological problems but also a penetrating re-examination of the human condition—body, mind, soul, and earthly home.

In writing about the emergence and synthesis of these new/old ideas, I have at times relied on my formal academic training (which concerns nineteenth-century literary forms and their connection with the science and culture of the day). But, more often, I have had to forge into unknown and widely diverse areas, reading ecological monologues, science textbooks, philosophical works, nature essays, novels, poetry, and various manifestoes (some serious, some flaky). At the same time, I have observed the ephemera of popular culture and counterculture, to see where experimentation seemed to be passing into wider acceptance. My goal has been to make sense of a complex and still-developing picture. I hope the reader can find half the enjoyment and excitement that I have felt, as I have begun to understand what has knit together the many, confusing threads of the ecological movement.

I have tried to walk the narrow path that runs between uncritical advocacy and dry detachment. These are exciting ideas. They deserve committed, sympathetic exposition. Yet they also need the kind of understanding that an informed historical perspective can give. Ecologism is not the beginning of a New Age: it is merely another episode in the long history that has shaped our minds. It will be, I think, a creative and fruitful episode. But probably it is not, as Charles Reich once wrote, the "new consciousness" that will overturn the ills of the West, nor the apocalyptic "turning point" that Fritjof Capra has announced. Let's keep things in perspective.

○ ○ ○

This book focusses on two crucial points in the process by which an idea makes its way into a role of cultural importance. One obvious approach is to look at sources. Where does the idea come from? This may be a historical exercise—every idea has a genealogy of some kind. Or it may be contemporary: when ideas are derived from science, they may have fairly clear points of recent origin.

Origins can be illuminating, but they don't tell all. They don't reveal why or where an idea is picked up from its hiding place in a scientific treatise or a forgotten philosophical controversy, and catapulted into public awareness. For this, one must turn to certain widely read or widely influential writings, seminal works that present the idea in a new, appealing, and relevant way, often in response to a particular public crisis.

So this book looks in two directions. It looks at the world of nature, through the eyes of ecologists and scientists who have, in a sense, newly minted it for us in a surprising and appealing way. These scientists would include Paul Ehrlich, Eugene Odum, Ramon Margalef, Ludwig von Bertalanffy, and Rachel Carson, along with A.G. Tansley, Frederic Clements, G.F. Gause, and Walter Cannon. And it looks at the world of ideas, through books and essays that have crystallized wide debate and diffuse emotion into powerful, focussed statement: among many others, the works of Lewis Thomas, Gary Snyder, Paul Shepard, Aldo Leopold, Wendell Berry, J.C. Smuts, and Alan Watts.

The implicit assumption of such an approach must be that movements of ideas get much of their strength from great and gifted men and women, around whose work the rest of us gather. They are the core, the head of the comet; and the popular mind—with its shifting fashions and ever-changing Gallup-poll readings—the periphery, the flashy but tenuous tail.

On the other hand, what makes a Carson or Leopold "influential," but influence? A chicken-and-egg problem vexes any attempt to study the flow of ideas, for without a certain public readiness even such brilliances as *Silent Spring* and *Sand County Almanac* would simply sputter away, unread and unremembered. Intellectual history attempts to gauge this readiness by assessing something maddeningly vague which is called a public mood, or a historical context, or a dominant paradigm. It amounts to making an imaginary model or construct, and saying: This is, in some kind of ideal average, how a significant group of people thought and felt at a certain time.

Most of this book concerns ecologism in its clearest and most memorable forms, the most reliable points within this construct: the scientific concepts that give it birth, and the ecological prophets who saw how those concepts could remake our intellectual and emotional world—our cosmos. Though much of this synthesis occurred in the 1960s and '70s, it had been prepared for by decades of work by both scientists and writers who had been touched by the strange, stirring news of ecology.

The Structure of Earth Rising

Though it is not a novel, this book has several plots.

One is the gradual replacement of a bloody, ruthless idea of nature—the Darwinistic view—by an idea that nature is serene, beautiful, and hospitable to life—the world as pictured by ecology.

Another is the struggle to find a single, unifying explanation for the two categories under which we habitually organize our world: mind and body, or spirit and matter. Ever since Descartes declared that these were two unlike and unrelated "substances," the question has floundered along in our intellectual history, oscillating from one side of the dilemma to the other. All is matter. Or no—all is really spirit. Or the two coexist, unrelated. But they *are,* obviously, related! And so on.

A particularly consequential form of this struggle has taken place within the science of ecology, which has had long traditions on both sides of the question, expressed as the materialist/reductionist approach and the holistic approach. Relying on the latter, ecologism offers new definitions of life, mind, and the material world that draw them into a single coherent reality.

A third plot in this book is the familiar conflict between science and religion. (It is easy to see how this relates to the previous dilemmas.) Does our factual knowledge of the physical world allow a place for spirit? For God? For love, or art, or the conscious, purposeful mind? Can we find a way to express our sense of value—what makes life worth living—that does not simply ignore what the sciences tell us are the facts of the real world?

I have tried to avoid oversimplifying these oppositions: good drama may be bad history. For example, on a popular level, Darwinian evolution plays the villain to the white-hatted ecology, which rides into town with good news of natural peace and beauty that vanquishes the bloody-minded ruthlessness of the Victorian-Darwinian jungle. In reality, however, ecology presupposes evolution: the ecosystem develops through the pressure of natural selection.

Hence, there is an additional plot here: the process by which scientists' ideas are adopted and used (scientists would often say "misused") in an attempt to fit them into a broader framework of human understanding. I am convinced that both sides of this process are important and respectable intellectual activities. The scientists are maintaining empirical rigor and skeptical inquiry; the framers of the worldview are trying to wrestle with the whole fact of existence, the inner world and the outer, to bring coherence out of it. Real people need both approaches.

○ ○ ○

In order to explore fully the worldview of ecology, this book places somewhat elementary introductory material in the first chapter, before beginning its more detailed analysis of ecological belief. Chapter 1 gives the "deep background" of ecologism: the culture of the middle ages and its overthrow by a new culture of science, technology, and materialism. Because ecologism must be seen as a new response to this basic condition, I have included this sketch of what may be, to some readers, very well-known facts of intellectual history. The first chapter also presents this material in the context of an equally basic (but essential) history of the rise of ecology and ecological thinking in the 1960s and '70s.

Readers who are already well grounded in these areas may wish to begin their reading at Chapter II. From that point, even those well versed in ecological literature will, I hope, find something of value in the attempt to correlate the scientific basis of ecologism with its beliefs, extrapolations, and habitual metaphors and parables.

CHAPTER I

Earthrise and the Readiness for Ecology

When an Apollo 8 crew member looked through the lens of his Hasselblad and photographed that majestic earthrise above the silvery edge of the moon on a December day in 1968, he created an image that penetrated the imagination of the world. The photograph was seen everywhere in the next years. Magazines loved it. It could be found on the cover of, or inside, numerous books. One major television network adopted it as the backdrop for its credits on the nightly news broadcast. In poster form that image decorated walls in bedrooms, offices, and college dorms throughout this country and abroad.

Merely as a photograph it is undeniably stunning. And no doubt it speaks many messages: about patriotism, technology, adventure. But one group of meanings speaks louder than any other. This picture vividly communicates the smallness, singleness, and frailty of our terrestrial home, and it does so without uttering a word. The shockingly unfamiliar image delivers not just an idea, but a breathtaking new perception of the nature of earth: a tiny green and blue capsule of life afloat in a vastness of serene unliving space.

The three astronauts aboard that vessel were officially engaged in preparing the way for an American landing on the moon, to take place within the next year. Their itinerary called for a figure eight loop around the back side of the moon and an immediate return to earth, and this they fulfilled to perfection. But they may have achieved incomparably more in simply being the first humans to travel far enough from earth to look back and view it as a single globe. They were the first to look at the back side of the moon. But what burned into the mind was the sight of the earth, alive with the swirl of clouds and the blue of sea, rising in bright contrast to the lifeless lunar horizon.

Frank Borman, commander of the mission, explained the experience this way.

> When you are privileged to view the earth from afar, when you can hold out your thumb and cover it with your thumbnail, you realize that we are really, all of us around the world, crew members on the space station Earth. Of all the accomplishments of technology, perhaps the most significant one was the picture of the Earth over the lunar horizon. If nothing else, it should impress our fellow man with the absolute fact that our environment is bounded, that our resources are limited, and that our life support system is a closed cycle. And, of course, when this space station Earth is viewed from 240,000 miles away, only its beauty, its minuteness, and its isolation in the blackness of space are apparent. A traveler from some far planet would not know that the size of the crew is already too large and threatening to expand, that the breathing system is rapidly becoming polluted, and that the water supply is in danger of contamination with everything from DDT to raw sewage. The only real recourse is for each of us to realize that the elements we have are not inexhaustible. We're all in the same space ship.[1]

Commander Borman's eyes were in a sense the eyes of the world when he looked out the spacecraft window—the world's *new* eyes, as if it were seeing itself for the first time. His thoughts were not those of a partisan American, winning a space-race trophy for his nation's pride. Nor were they those of a cocksure military engineer-pilot, flying the hottest piece of technology ever made. Rather unexpectedly, his thoughts were those of a world-citizen and ecologist. His firsthand interpretation of this sight identified exactly those themes that were becoming the heart of the ecological worldview, down on planet earth below.

What are those themes? First Borman notices the homeyness and smallness of earth, the qualities that awaken not the desire to conquer, but the sense of belonging. This is partly the voice of the traveller, a long way from home. But it is also the recognition of a fundamental loyalty and affection for the entire planet and all its entangled systems of life.

Next Borman emphasizes the oneness of the earth: from this perspective, humankind's differences are insignificant. The biosphere defines humanity's true boundaries, not the political atlas or the book of ideology. This leads to the main point of the reverie: the unity of the earth is also the limitedness of the earth. Earth is "bounded," "limited," a "closed cycle."

This new perspective on humankind's home is the heart of this picture and its power. The feelings and ideas Borman expressed on this occasion were the beginnings of a revolution in thinking which, over the next fifteen years or so, would deeply alter the way individuals and societies thought about themselves and their world. The impact of this new perspective is comparable to other historical shifts in human thinking caused by dramatic changes in viewpoint. When explorers who followed Columbus brought home the startling news that there was a whole new continent sharing the planet with the Europeans, it contributed to a reawakening of the European mind, a burgeoning sense of the world's bigness and possibility. Likewise, when Copernicus shifted the earth from the center of the solar system, he effectively moved heaven and earth. These discoveries caused a profound rethinking of the human place in the cosmos, and were no small part of the general awakening of the Renaissance.

The ecological revolution does the same thing, and perhaps even more radically. For this new perspective challenges virtually every basic tenet of belief about human identity, the shape of society, and the nature of the world. And it demands not just new conclusions but a whole new style of thinking.

An Unexpected Manifesto

Probably the most persuasive evidence of the widespread influence of ecological thinking in the U.S.—though for most people it was undoubtedly unconscious, unsystematic, and unexamined—lies in the massive social and political support given to environmental legislation in the late 1960s and early 1970s. Within a few years, early successes were on the books: the National Environmental Protection Act (1969), the Air Quality Act (1967, with important revisions in 1970), and the Federal Water Quality Act (1972).

Less than two years after the Apollo 8 venture, even the old Cold War Republican then in the White House found it desirable to get on the environmental bandwagon. Having already appointed the Environmental Quality Council, Richard Nixon went on to enthusiastically support the Environmental Protection Agency (established by executive order in 1970), climaxing years of rising concern, popular agitation, and piecemeal legislation on state and federal levels.

President Nixon called the environment "the issue of the decade," and produced an astonishing prologue to his Council's report (August 1970) in which nearly every aspect of ecological thinking is presented and advocated. Here is an excerpt:

> The basic causes of our environmental troubles are complex and deeply embedded. They include: our past tendency to emphasize quantitative growth at the expense of qualitative growth; the failure of our economy to provide full accounting for the social costs of environmental pollution; the failure to take environmental factors into account as a normal and necessary part of our planning and decision-making; the inadequacy of our institutions for dealing with problems that cut across traditional political boundaries; our dependence on conveniences, without regard for their impact on the environment; and more fundamentally, our failure to perceive the environment as a totality and to understand and to recognize the fundamental interdependence of all its parts, including man himself. . . . We need new knowledge, new perceptions, new attitudes. . . . We seek nothing less than a basic reform in the way our society looks at problems and makes decisions.[2]

I quote this at length because it neatly sums up so much of ecological thinking. If taken seriously, this is a manifesto for revolution. That it comes from a very conservative politician sitting atop the hierarchy of the world's greatest accumulation of industrial and military might is practically flabbergasting. It suggests that the President did not always read, or perhaps believe, everything he set his signature to;

but also that there was a very widespread public readiness to question old ways of thinking, and consider the alternatives offered by the ecologists. These alternatives ran clean against the established American values of rugged individualism, technological progress, and unrestricted competition. In their place, the ecologists offered a sophisticated program of human and natural interdependence, of limits to growth and the need for a stable society, and of the moral imperative for communal cooperation. These are the ideas of the ecological worldview.

The Fragmented World

What kind of a world was it, into which this ecological way of seeing burst with such suddenness and power? In terms of intellectual history, the conditions had long been ripe for some such unifying vision.

For the ecological worldview has arisen in a fragmented modern world. The fragments are, to a large extent, the pieces left over from what might be called the "old order": a comprehensive philosophical and religious system based on the traditional conviction that the world is as it is because the omnipotent God of the Bible made it that way. As a system capable of explaining the totality of life, the old order began crumbling long ago, though modern languages, institutions, and religions still carry elements of it (some of them still quite influential).

The old order, often summarized in the phrase "the Great Chain of Being," was a world-picture that lasted in various forms from the Middle Ages until at least the end of the Enlightenment. It depicted a continuous hierarchy, created by God, that descended by orderly links from heaven to earth; from the Trinity to the Pope to kings, subjects, and serfs; from the crystalline spheres of the planets and stars to inanimate matter on earth. It was a static and hierarchical structure within which everything had its correct place, securely above or below its near neighbors. For a long time, the Ptolemaic (geocentric) universe gave a good representation of this order, placing the earth in the center according to its importance, and arranging all else in perfect circles around it. Biblical authority supported the works of philosophers and astronomers who reasoned about and explored this world; all was bound into a secure, cohesive order, with God on His throne, humankind His principal concern, and the earth the center of the universe as the theater of human salvation.

We are far from that beginning, yet also near. Though of course religious belief persists, it has been removed from the center of Western culture. It no longer provides the basic framework for our understanding of astronomy, biology, psychology, law, ethics, and so on. Even the most orthodox religious believers today are likely to think in modes quite alien to the ancient tradition: a scientific sense of fact, a historical sense of time, a psychological sense of the mind. But in a very real sense, our Western culture is still struggling with the breakup of this old orderliness. In the midst of rapidly changing social conditions, we find our moral bearings almost as unreliable. The serene confidence of the old order has been replaced with a confusion of private answers and partial solutions, and a society-wide sense of existential uncertainty.

❍ ❍ ❍

The story of how science and rational skepticism overturned the old order is the backdrop for the story of ecologism. The familiar reference points are René Descartes (1596-1650) and Isaac Newton (1642-1727). Descartes' division of the world into two unrelated substances, mind and matter, began a centuries-long process of dividing the facts of the material world from the values of the mind and heart. Newton's pursuit of this Cartesian idea of a machine-like material world resulted in perhaps the single most stunning advance of science in history: the mathematical understanding of the motions of the planets. In fact, Newton lent his authority to the strict materialist and mathematical understanding of *all* physical phenomena. The world was to be understood not by reference to the Divine Will, but by reference to natural law, operating at the fundamental level of the atom itself:

> God in the Beginning form'd Matter in solid, massy, hard, impenetrable, moveable Particles, of such Sizes and Figures, and with such other Properties, and such Proportion to Space, as most conduced to the end for which he form'd them.[3]

The implication of this view is all too evident. Though divine purpose is invoked, Newton effectively limits God's existence to that of a premise at the beginning instant of the universe—a "first cause." After that first instant the laws of nature are sufficient to govern the interactions of the massy, movable particles. Indeed, the entire history of the universe would seem to be implied in the original disposition of those particles. Newton saves the idea of Providence by referring to the physically determined outcome as "the end for which he [God] form'd" matter—but this was a bit of piety which did not really affect anything in the system. It was an extra assumption which could be, and would be, dropped from the logic of science with Laplace's famous "I have no need of that hypothesis!"

Descartes and Newton were both solid religious believers, who could hardly have foreseen the results of their work. Nevertheless, the science which they helped create became the cornerstone of a thorough application of reason and skepticism to all parts of the human and natural world. By the end of the eighteenth century, the "method of doubt" was being applied to the foundations of religion and even the logical operations of the mind itself, most notably by David Hume and Immanuel Kant. They showed that the two pillars of the old order, revealed religion and human reason, were as mortal, fallible, and changeable as anything else. Thus, somewhat shockingly, the Age of Reason ended in a flurry of radical doubt about its own method. The confident rationalism with which the enlightened Deists had replaced traditional faith had proved to be a self-destructing artifact.

The contemporary Western world still struggles with this legacy. No enduring sense of the cosmos has arisen to replace that demolished by reason, science, and skepticism. The success of science has put moderns in exquisite control of the physical world, but left them inwardly at a loss, orphaned and alienated and incapable (save by heroic individual efforts) of drawing their existence into one clear meaning. Men and women since the eighteenth century have faced an almost unbridgeable gulf between their inward beliefs and their sense of the realities of the outer world. In the words of a fine analyst of the modern condition, Robert Langbaum, this has created a "wilderness" of spiritual wandering and moral dismay.

> That wilderness is the legacy . . . of the scientific and critical effort of the Enlightenment which, in its desire to separate fact from the values of a crumbling tradition, separated fact from all values—bequeathing a world in which fact is measurable quantity while value is man-made and illusory.[4]

Against this backdrop of cultural breakup and change have occurred numerous attempts not to oppose science (a seeming impossibility), but *to use it to reconstruct a livable world order.* If the value-organized world of the Great Chain had crumbled, then a fact-organized world-view would have to replace it. Such values as there may be would have to be consistent with science, and ideally—if possible—derived from it.

When Newtonian science came to be applied to biology in the form of Darwin's evolution by natural selection, the worldview called Social Darwinism soon emerged to try to make sense of it, to lay out the moral and ethical structure of a world so defined. It was a hard task, to make an acceptable ethics out of the ruthless logic of the "survival of the fittest."[5] Yet many tried: most notably Thomas H. Huxley and Herbert Spencer in England; William Graham Sumner in America; many eager industrialists in both countries, ready to wed capitalism to Darwinism; and even scientists like Darwin and his codiscoverer of natural selection, Alfred Russel Wallace. The world-view they created tried to make a virtue of necessity by frankly embracing nature's harshness: letting the weak suffer and perish was the law of nature, but it was, after all, also the "law of progress." Even before Darwin's *Origin of Species* (1859), Herbert Spencer was discovering a kind of natural selection at work in nature as in society:

> Pervading all nature we may see at work a stern discipline, which is a little cruel that it may be very kind. That state of universal warfare maintained through the lower creation, to the great perplexity of many worthy people, is at bottom the most merciful provision. . . . The poverty of the incapable, the distresses that come upon the imprudent, the starvation of the idle, and those shoulderings aside of the weak by the strong, which leave so many 'in shallows and in miseries,' are the decrees of a large, far-seeing benevolence.[6]

Darwin's theory was regarded as providing a scientific basis for such views. Referring strictly to the natural world, Darwin wrote: "Thus, from the war of nature, from famine and death, the most exalted object which we are capable of conceiving, namely, the production of the higher animals, directly follows."[7] Many concluded (unscientifically) that this natural law was also human destiny. In human society as in nature, struggle, competition, and widespread suffering simply had to be accepted. By the suffering of the many, the species would continue to advance. In the cheerful words of American steel magnate Andrew Carnegie: "All is well since all grows better."[8]

The assumptions of this scientific worldview were those of nineteenth-century positivist science: materialism and atomism. From Descartes, it had inherited the dichotomy of two substances, mind and matter; from Newtonian science, it had concluded that only the "matter" side was real (the world is "made of ether and atoms, and there is no room for ghosts"[9]), and that the fundamental reality was a law-governed, billiard-ball-like "concourse of atoms." As Huxley put it:

> The whole world, living and not living, is the result of the mutual interaction, according to definite laws, of the forces possessed by the molecules of which the primitive nebulousity of the universe was composed.[10]

The net result of this worldview was to reduce human life in all its mental, emotional, and spiritual variety to mere chemical effects on one level, and survival of the fittest on another.

By the early decades of the twentieth century, the power of this grim worldview had begun to diminish rapidly. Darwinian science developed and prospered; but the expectation that this new science could provide a master key to life withered away and disappeared. Simply put, this approach could not make a coherent and emotionally satisfying explanation for the whole of life. It left out too much—love, altruism, divinity, the human spirit. Indeed, Darwin's ruthless and bloody picture of nature exacerbated the already intense modern dilemma. It made the facts of the world seem further than ever from the values of the human heart. Both Huxley and Wallace eventually came to renounce the worldview of Social Darwinism, both in fact hitting upon the word "nightmare" to describe its emotional effect.[11] Of course they strongly continued to uphold the scientific validity of evolution by natural selection. But they began to look elsewhere for a sense of the whole.

In this context, the emergence of ecologism is interesting indeed. Like Social Darwinism, ecologism aims to be a *scientific* worldview: it extrapolates its ideas from a scientific description of the natural world. Thus, "nature" is taken as the ground and beginning-point; science (though sometimes imperfectly represented) is incorporated near the foundation of the worldview; and all the higher conclusions are, in theory at least, consistent with the physical realities. Such respect for science and the physical world is, no doubt, a necessity for any successful worldview in the contemporary world. Both ecologism and Social Darwinism illustrate this generalization.

But the worldview of ecology, working from immensely changed and broadened biological materials, is able to picture a surprisingly hopeful cosmos. For it draws from scientific ecology a wealth of suggestive ideas about the cooperativeness, harmony, creativity, and intelligence of nature, offering a striking antidote to the bloodiness of Darwin's natural jungle, the coldness of Newton's mechanical universe, and the fragmentation of Descartes' two disconnected substances.

The "Earth Household"

Ecology, as is well known, is a word coined out of the root for "home" (Greek *oikos*). It is the science that studies the interactions of the living and nonliving components that together make an ecosystem. Ecology describes how nature manages its household—how the varied requirements of life are met to such satisfaction, by so many creatures, in so finite a space. The word also suggests the powerful emotional content inherent in that scientific description. The world as ecology sees it is indeed a homey place, a realm of intricate and marvelous arrangements all made to accommodate life. This is an emotional extrapolation, of course, and an oversimplification: ecology does not ignore death and suffering. But it contains these realities within an overall framework of thriving, mutually beneficial life. The ecological worldview is powered by this feeling, this intellectual and emotional grasp of the friendliness of the biosphere.

Ecology is most interested not in isolated creatures, but in how they connect to make larger groups and units. Even when the ecologist looks at a single species or individual, he or she is interested in how it fits in to the whole, how it makes its living or finds its niche, what its relation is to other creatures and influences. This leads ecology to see living things as essentially *connected.* No plant or animal could survive or exist without the whole apparatus of nature supporting it and its unique role. The individual is seen in ever-wider contexts: its population, its species, its ecosystem, its biosphere. Ecology, practically alone of all the sciences, has historically recognized the reality of these ever-widening circles of connectedness. Hence ecology also came to be, as Eugene Odum's 1975 textbook puts it, "a major interdisciplinary science that links together the biological, physical, and social sciences."[12] Ultimately these contexts all form one unit, of which every creature is a fully integrated part.

For the human animal, it is a point of view which creates an unexpectedly poignant sense of belonging. Earth and all its creatures form one biosphere.

Along with this affecting sense of earth's unity, however, comes the sobering sense of how finite it is, and how overloaded are its finite systems. The ecology movement was not born out of a photo. It was galvanized into life in response to enormous problems: overpopulation, pollution, and resource scarcity. All were problems of finitude, limitedness—that circular self-containedness so vivid in the photograph. The population could not expand forever without encountering the natural ceiling of the biosphere's carrying capacity. And pollution could not continue to be dumped indefinitely into the atmosphere and waters, without degrading their ability to support life. The living earth, though thriving, was also fragile.

These were new problems in the sense that it took a changed worldview to perceive them fully. The dominant attitude in the West has been belief in "progress." This is a highly linear view of time, one which emphasizes ceaseless growth, change, and improvement. At best, earth is merely the theater for this unfolding of human potential. At worst, it is an obstacle to it. To acknowledge limits imposed by the planetary biosystem would be, from this perspective, to admit defeat—to abandon humanity's destiny.

Socially and politically, this attitude was powerfully expressed in the American pioneer. The pioneer attitude was, of course, to chop, plow, mine, exploit, and conquer everything in its path. The idea that its very path might come to an end—at the edge of a continent, for instance—or that its exploitable resources might become exhausted, was simply not a part of its way of seeing the world.

The eruption of pollution and population problems in the 1960s and 1970s (along with the corollary of overpopulation, resource scarcity) has begun changing this perspective, however. The open-ended world of American pioneer aspirations has been abruptly challenged by recognition that there is only so much room for people to live in, only so much clean water to drink, only so much oil to pump out of the earth. The linear, endless time line is contradicted by the ecological images of limited space: the closed circle of the earthly globe.

Crisis and Response

One of the first books to present an ecological view of the crisis was Paul Ehrlich's *The Population Bomb* (published in repeated editions from 1968), which told in scientific and global terms what most people had been able to observe for themselves, as the postwar baby boom gave local impact to the abstract statistics of worldwide population increase. Ehrlich is an eminent ecologist. It is the most basic of ecological tasks to determine how a population either grows, stabilizes, or dies off, in interaction with the ecosystem that carries it. His book vividly presented the implications of a doubling or tripling world population.[13]

An equally influential book was *The Limits to Growth,* published in 1972 by a radically earth-centered, transpolitical group, The Club of Rome. Its members described themselves as "approximately seventy persons of twenty-five nationalities, . . . united . . . by their overriding conviction that the major problems facing mankind are of such complexity and are so interrelated that traditional institutions and policies are no longer able to cope with them, nor even to come to grips with their full content."[14] According to *The Limits to Growth,* humankind was indeed united in its reliance on a single, finite earth. The problems facing humanity were insoluble by traditional ways of thinking, familiar ideologies, and national concerns. They were problems for an earth-centered point of view.

In fact, these problems forced a re-evaluation of the most basic premise of modern civilization: unlimited growth. The book assembled the grave data of "world population, industrialization, pollution, food production, and resource depletion," and concluded that current practices could not go on much longer—probably less than a hundred years. Collapse would follow. The only alternative to uncontrolled growth in all these areas was, therefore, "a condition of ecological and economic stability that is sustainable far into the future"—a "global equilibrium."[15] As the chapters that follow will detail, the notion of equilibrium is a most basic ecological idea; it is the foundation for nature's self-management. And humankind would have to learn it.

Coming hard on the heels of *The Limits to Growth* was an event which drove home the reality of the world's finitude. The Arab oil embargo of 1973-74, and the resultant "energy crisis," told Americans and, in fact, the entire industrialized world that the problem of limits could not be ignored or deferred. It was already upon them.

Ecological scientists stepped into the midst of this clamor of problems with explanations and solutions. It made them controversial—

but they were listened to. The interlocked problems of population, pollution, and resource scarcity were comprehensible only within the ecological framework. They required one to think in terms of whole systems, and their finite ability to sustain life. They began to force the ultimate change in perspective: the grudging acceptance of the notion that humans were inescapably subject to the laws and limits of biological nature.

Despite the brief fame of Social Darwinism—itself never more than a minority worldview—it was still to most people a rather new idea that humans had a purely natural identity. It was still an odd (if fascinating) new thought, that trees, plants, animals, and people all were indivisibly locked into the various webs of life. One couldn't seem to poison a bug without poisoning lots of other things—streams, lakes, fish, birds, people. To pollute a stream was to destroy whole systems of life, in branching webs that eventually touched human beings. Even a decision about the number of children to have had repercussions far beyond one's own family. Accustomed to seeing themselves, and other organisms, as individual units, Americans and others found this new ecological picture of life rather startling. For it bound all creatures into a single system, an ecosphere, from which there was no "independence." But if startling, it was also hard to shrug off. For the stresses industrial civilization had begun placing on the environment had become all too obvious and frightening.

The Environmental Focal Point

The problems of environmental pollution described by scientific ecology exploded into the national mind and political agenda at the end of the 1960s. It was a time of already overcharged national emotions: the Viet Nam war raged at its height, and the civil rights movement still claimed preemptive attention from many Americans. Nevertheless, when the first Earthday celebration occurred in April 1970, the public, the media, and the government were ready to be galvanized into seemingly instantaneous action. "Teach-ins" proliferated. The process of agitating for legislative action accelerated. The

President had recently appointed the Council on Environmental Quality. Students added the environment to their list of primary concerns. And "ecology" swiftly began to accumulate meanings that took it far beyond its scientific basis.

Barry Commoner, the biologist/activist who was quickly becoming the new movement's best-known spokesperson, wrote during that year:

> The sudden public concern with the environment has taken many people by surprise. After all, garbage, foul air, putrid water, and mindless noise are nothing new; the sights, smells, and sounds of pollution have become an accustomed burden of life. To be sure, the mess has worsened and spread in the last decade, but not at a rate to match the dramatic, nearly universal reaction to it that has hit the country in the past year.[16]

Commoner goes on to suggest that the sudden popularity of ecology occurred not in spite of the other pressing national issues, but *because* of them. In some way the environmental problem seemed to encapsulate all that was wrong with modern life: its urban artificiality; its industrialization; its dehumanizing focus on profits and power; the war abroad; racism at home. These problems seemed to make a package of traditional values gone terribly wrong. Conversely, a thoroughgoing ethic of life-affirmation tried to incorporate concern with the biosphere into its struggle against warmaking and racial injustice. Solidarity with one's fellow humans merged into solidarity with all life.

❍ ❍ ❍

In its early days, the ecological movement was most visible in the extreme forms embodied by countercultural radicals, who amalgamated a surprisingly diverse array of values and images borrowed from the avant-garde, pop and rock music, and non-Western cultures of almost any description. One of the foremost representatives of the movement was (and is) the ecological poet Gary Snyder, who addressed his 1969 book *Earth House Hold* "To Fellow Dharma Revolutionaries." He embodies both the seriousness and the eclecticism of the movement; he spent years as a monk in a Zen monastery in Japan, and sees himself as part of "a tradition with roots in the paleolithic." He correlates his life and work with many other sources as well: American Indian song and myth, the Tao, the *I Ching,* Yoga, Shamanism, and so on.[17] For the movement at large—less disciplined and informed than Snyder, but fully as eager—almost anything could be accepted as a part of the "New Age" or "Age of Aquarius" if it talked about renewal or rebirth or seemed to promote unity, peace, nature, or freedom; or if it opposed technology, industrialization, progress, or warfare. Charles Reich points out how important the idea of naturalness was to this counterculture: the new science of ecology had defined for it a new kind of nature, full of cooperativeness and accommodation, which provided an acceptable basis for its antiwar, reformist agenda.[18]

The emergence of the Green Party in West Germany, like related Green and Bioregionalist movements in the U.S., illustrates that Commoner's 1970 observation still holds true: the environmental movement in a fundamental way sums up a whole range of issues and attitudes which challenge the traditional structure. Petra Kelly, an early leader of the Greens, expresses the linkage this way:

> I like to call [the Greens] a kind of planetary spiritual movement because we try to see interconnectedness. For example, the arms race, the unemployment rate, the inflation, the export of arms to the Third World, the export of nuclear power plants, it is all connected. As one of the astronauts has said here in the United States, if you look at the planet, you see no boundaries, you see no states, you see just a flowing Earth, a planet on which we are all together.[19]

The party is "green" because its assumptions about the world are ecological. It believes earthly unity to be a *fact,* however obscured by habits of thought and traditions of politics. It believes cooperation to be a powerful and neglected survival tool. It sees warmaking and pollution as merely two expressions of a human spirit isolated from nature, and grown vicious and arrogant as a result.

The centripetal force of ecologism—its ability to attract and fuse a diverse array of social, political, ethical, and spiritual concerns—

soon began to be seen in another area: the emergence of ecofeminism. The affinity of feminist and ecologistic visions is not hard to see. The feminist movement offers a deep critique of paternalistic values such as aggressive individualism, hierarchical social structures, domination of others, warmaking, industrialism, emphasis on abstract ideology at the expense of the needs of real people, and the like. The term "ecofeminism" was coined in 1974 by the French writer Francoise d'Eaubonne: as Mary Daly remarks in her 1978 book *Gyn/Ecology,*

> [d'Eaubonne] maintains that the fate of the human species and of the planet is at stake, and that no male-led "revolution" will counteract the horrors of overpopulation and destruction of natural resources. I share this basic premise.[20]

Historically, Western culture has identified nature with the female, and conversely woman with nature, a fact to which the ecofeminist movement has reacted variously. As an instrument of oppression by male-dominated culture, this woman-nature identification has been rejected energetically by some feminists. However, others have celebrated the tradition of earth as nurturing Mother through a restored appreciation and awareness of ancient goddess worship and cultus.

This historic identification underscores perhaps the most basic reason for ecologism and feminism to come together. Ariel Sallah, a feminist who became a leading Australian Green leader and ecofeminist, describes the connection this way:

> I realized it was the same story. The oppression and exploitation of women was the same, was based on the same attitudes, as the oppression and exploitation of nature.[21]

But ecofeminism does not merely represent two groups making common cause in the manner of a political alliance. Their connection is essential. Ecologism and feminism largely agree that reacquisition of what have traditionally been designated feminine styles of thinking and being are the very means by which environmental sanity can be restored.[22] The paternalistic, exploitative, individualist, and abstractly ideological habits of Western culture must be replaced by a nurturing, cooperative, communal way of being that is rooted in real connectedness to other humans and the earth.[23] Typical Greens statements of key values or points of unity include a strong ecofeminist declaration,[24] as do most serious works of ecologism. The exact relations of women to nature and to the transformation of masculine industrial culture remain central and fruitful questions in both movements.

○ ○ ○

In its early days, the young followed this developing ecological revolution in vast numbers. Their well-known demographic clout—over half the nation's population was under thirty in 1972—made their fads America's fashions. And the outdoors quickly became the campus favorite, as students readily added to their already-popular blue jeans items of mountaineering gear from the quickly proliferating specialty stores and catalogues: down parkas, huge red-laced boots, knapsacks with prestige makers' labels prominently displayed. One might have thought, in looking over a college campus on a coolish day, that everyone was just back from the wilderness. In fact, as the effects of the baby boom and the emphasis on nature combined, use of the National Parks skyrocketed, more than doubling in the decade of the sixties, and more than tripling from 1955 to 1974.[25]

A widespread impact was felt in the area of health. As Rachel Carson has remarked, "There is also an ecology of the world within our bodies."[26] Ecological thinking stressed the need to respect the natural system, to allow its inherent stability and productivity to thrive. The ideal would be to "design with nature," in Ian McHarg's phrase, rather than to force human designs upon it. Inherent in this notion was a new definition of health—one that could apply equally well to the human body or to the land. Health, from this perspective, was not the absence of disease; health was what happened when a complexly balanced living system ran smoothly. Hence, the correct way to treat disease—like the correct way to treat ecological problems of erosion or tree blight—would not be simply to attack the immediate cause, but to restore the system's depressed ability to resist. The pesticide and the bulldozer, the chemical medicine and the scalpel: these only removed the symptom of ill health, not its cause. Increased awareness of healthy foods, vitamins, and exercise reflected a distinctive ecological contribution.

The social consensus created by environmentalism by the mid-seventies was not limited to recreation and brown rice. For example, consider the attitudes of a group of major-corporation executives gathered under the auspices of "The Conference Board" in 1974-75. for the purpose of discussing current issues. Their clear support for environmental goals is quite startling, especially considering the very frank and energetic tone of the meetings. "EPA is an absolute necessity," according to one. Most of those attending supported "federal standards of environmental quality," and strongly criticized "companies which pollute our water and air and are apparently indifferent to the hazards of pollution until the Government arrives."[27] Similarly, a 1974 survey of 325 students at a western university found *no* business student expressing disapproval of the environmental movement itself.[28]

With support even from conservative elements, it is hardly surprising that important public policy decisions took decidedly ecological turns. Development of a technological wonder, the Supersonic Transport, was stopped cold after millions of dollars of research, primarily on grounds of environmental concern. There was a sense that the convenience and prestige of the extra miles-per-hour simply did not repay society for its expenditure of natural and social resources. An amazing development, less than two years after the first moon landing! This willingness to forego technological "progress" out of concern for the net effects on society and the environment was followed by a parallel social choice against nuclear power—a choice made in the midst of huge energy-cost increases and general economic distress. Eventually, the accidents at Three Mile Island and Chernobyl would solidify the sentiment against nuclear technology, and virtually stop construction of new nuclear generating plants. These choices betoken a new way of looking at technology and new values concerning progress and nature. These new values now compete with still-entrenched older ones, and constitute an important set of alternative answers on many national issues. They are the new way of thinking once called for by President Richard Nixon.

A truly new way of thinking is a tall order, even in as dynamic a society as ours. Yet in the ferment of the late sixties and early seventies, with ecological crisis demanding attention, a bloody war illustrating failure, and a stirring picture from space calling out yearning for success, pieces of a worldview that unified the fractured earth began falling into place. Some of these pieces were newly discovered by science; some were feelings and expressions that had been with us a long time. Both practical and moral, political and quasi-religious, the new worldview sprang into being, it seemed to the average citizen, overnight.

○ ○ ○

The chapters which follow look in detail at the elements of this worldview which aims to reunite fact and value, knowing and feeling, science and religion, humanity and nature.

It may be worth pausing here, however, to observe that none of the three worldviews mentioned so far—the Great Chain of Being, Social Darwinism, or ecologism—is ever perfectly successful at unifying all the diverse threads of existence. Not the least reason for this imperfection is the sheer immensity of the worldview edifice: it embraces too much of everything not to include plenty of inconsistencies. And while change may be swift in knowledge about the physical world, only gradually do the implications make their way into the broader moral and intellectual structure. All its parts interconnected, it is not quickly or easily reorganized.

The Ptolemaic (geocentric) picture of the cosmos, for instance, was often referred to in general culture by medieval and Renaissance theologians and literary writers as a convincing illustration of the orderliness of God's creation. Yet, at the same time, the astronomers who were trying to actually *use* this system were subjecting it to all manner of refinements, elaborations, and distortions to get it to match the observed phenomena (specifically, the oddball forward-and-backward motions of the planets against the more stable backdrop of the stars). These complications rarely attracted the notice of nonspecialists, and did nothing to weaken the importance of this picture as the emblem of universal order.

In the same way, many of the scientific concepts on which ecologism rests are subject to question. Contradictory data are available for concepts such as the climax ecosystem and its connection to diversity, for instance. Similarly, while ecologism relies on a holistic vision of nature, many ecologists are turning away from the holistic approach and returning to a more traditional mechanistic style of science. Ecology, in fact, has always struggled with an inner tension—almost a contradiction—between the kind of science practiced by physicists, and some other kind that seems more suited to biology. The recent drift in *scientific* ecology back toward more traditional mechanistic concepts has not immediately affected the ecological *movement*—for, on most topics, the debate is still open.

Despite this current re-examination of its own methods, ecological science has provided a set of new ideas with radical implications that have never been hinted at by solid old Newtonian physics with its predisposition toward the atom, the individual, and a few laws of motion and thermodynamics. There is a continuing willingness to consider the importance of the larger units in biology—the whole organism rather than merely its chemistry; the whole species rather than strictly its individual members; the whole ecosystem rather than only its component species. Further, the science of biology is facing up to the ability of living things to gain in complexity and to hold off the usual thermodynamic decline. In ways like these, the science of ecology is adding a whole new layer of ideas to the relatively simple (and disturbing) ones laid down by the tradition of Descartes, Newton, and Darwin. These, as we shall see in chapters to come, are being picked up and elaborated into a striking and impressive new way of looking at life on earth.

And if it is true that some parts of the ecologistic picture are being questioned by scientists, this is only to be expected. The more interesting question is, whether the ecological picture of interconnectedness and harmony will prove to be a temporary fancy, or—as seems likely—one of the most enduring products of the twentieth century.

CHAPTER II

Holism

The new view of the earth as a whole achieved by Commander Borman and his crew loosed a new awareness upon the world. The Apollo 8 "Earthrise" photo is the best known agent of this new awareness; but to a perceptive few the message had been clear much earlier. A few months before the first men orbited the earth, Loren Eiseley observed how, from the perspective of space, humanity's vaunted superiority and mastery over nature would take on quite a different look:

> Man in space is enabled to look upon the distant earth, a celestial orb, a revolving sphere. He sees it to be green, from the verdure on the land, algae greening the oceans, a green celestial fruit. Looking closely at the earth, he perceives blotches, black, brown, gray and from these extend dynamic tentacles upon the green epidermis. These blemishes he recognizes as the cities and works of man and asks, "Is man a planetary disease?"[1]

The abiding question implied in this vision is: Does humankind belong? Is it a part of this vast organism, or merely a dreadful and extraneous blight upon it?

These reflections, like Commander Borman's concerning "Spaceship Earth," are produced by a new way of seeing—in effect, a new reality. This is a reality in which the divisions and differences that loom so impressively from close up are regarded, from a different perspective, as minor details within a larger unit or system. The habit of seeing the "whole" which various parts make up, instead of focussing on this or that component, is a hallmark of scientific ecology. As a general approach to thought, it is called *holism.*

○ ○ ○

As a formal method of understanding, looking at things in the whole, rather than analyzing them into constituent particles, is a tradition of very long standing, reaching far back before the Cartesian/scientific method, and continuing (as an alternative or minority stream) into the present day. It enjoyed some success in the seventeenth and eighteenth centuries as an approach to biology and chemistry, even while these disciplines were for the most part developing towards a strictly analytical method. But this way of seeing did not receive its familiar name until 1926, when it was formally dubbed "holism" in a widely noticed book by Jan Christiaan Smuts.[2]

He was a curious man: a general and political leader of South Africa during formative years of the 1920s, his name still adorns the airport of the capital city of Johannesburg. Smuts had the honor of imprisoning the young Gandhi for civil disruption, and of fathering a modern state explicitly based on racism. Hardly the sort one would expect as a grandsire of radical ecologistic thinking!

Yet obviously his interests and talents ranged far beyond the borders of his troubled state. The word he coined, and the integrated style of thought he advanced, have re-entered Western thinking as revolutionary forces of change. Coming when it did, his presentation of holism contributed a handy catchword and a readable exposition to the ongoing debate about wholes and parts—a debate in which the just-emerged science of ecology was deeply embroiled.

General Smuts' *Holism and Evolution* is still read for its forceful presentation of the essential ecological notion that the world is best understood in its relations. It works from an underlying intuition that the various wholes of our experience—our bodies, our minds, our societies—are not accurately represented by reductive analysis. Smuts understands his task very clearly: to get beyond mechanism and atomism. He attacks the assumption that "there could be nothing more in the effect than was already in the cause," which inevitably means that mind and "soul" could only be "mere shadows or unreal accompaniments of some real mechanical process."[3]

Smuts attempts to work from a rather lofty vantage point, combining relativity theory and evolution:

> Just as a thing is really a synthesized event in the system of relativity, so an organism is really a unified, synthesized section of history.[4]

He proposes to see reality in terms of "fields" rather than "things." Time has borne out his eager acceptance of the new model of the atom as a set of interacting energy fields rather than a simple unit of inert matter. From the atom upwards, the real world for Smuts comprises complex systems, or "real wholes," at multiple, interlocking levels. Each level is real in itself, but also forms part of a larger reality. A real whole is typified by "a unity of parts which is so close and intense as to be more than the sum of its parts."[5]

Smuts takes special notice of ecology, which at this time was well into its classical early stage, represented by the works of Frederic Clements and Charles Elton. Smuts admires "the young science of ecology" for its revision of Darwinism to include a "friendly" version of natural selection.[6] This would become an important part of the total package of ecologism: the insight that, if the system of nature were regarded as a whole, this world would be seen as a beautifully hospitable place for life to unfold. With a keen eye for the telling metaphor (demonstrated many places in this book), the Afrikaaner borrows the powerful ecological picture of a unified and welcoming nature. So potent did this picture seem to him that he uses it as the concluding image for the entire book. Smuts declares that humankind lives amidst

> the grand Ecology of the universe. ... the *oikos,* the Home of all the family of the universe, with something profoundly friendly and intimate in its atmosphere. . .[7]

Smuts' work suggestively presented ecology's connection to the problem of parts and wholes, and gave impetus to the movement away from reductive scientific atomism. But the General's little book was both followed and preceded by more weighty works on the subject. The philosophers Bertrand Russell, Samuel Alexander, and

Alfred North Whitehead were all at that time developing ways of logically describing "real wholes." The question is an ontological one: a "real whole" has some inherent quality that cannot be attributed to its parts, as Smuts insisted. In this sense an organism may be real *qua* organism—not merely real as a quantity of matter. In Russell's terminology, it is of a different order of logical type. Of course, there has been a powerful scientific bias against this kind of approach, traditional science preferring analysis of the constituent parts to understanding of the integrated whole.

The Oneness of Nature

"In nature nothing exists alone." These words by biologist Rachel Carson echo the earlier motto of John Muir: "Everything is connected to everything else." Both statements efficiently summarize the ecological viewpoint. The trick of seeing things separately—say, an individual tree abstracted from its forest and climate—is as artificial and misleading as looking at a river without reference to the region it drains. Neither trees nor rivers exist in isolation. In the dimension of time, a tree's biological history connects it to perhaps millions of other individuals. And in the dimensions of space, a tree both affects and is affected by hordes of other creatures, from the microscopic life in its roots to the birds and animals in its branches, from the squirrel that buried its original seed, to the plants which for some reason left a place for it to grow, to the termites and beetles and saprophytes that will consume it when it dies. An "individual" tree, in this tradition of ecology, is in some significant ways less meaningful than the tree seen as an embedded aspect of a history and an ecosystem. The life of that tree, its pattern of growth and reproduction and death, its very chemical molecules, are all inextricably bound up with other lives and with the physical characteristics of its place.

Thus the science of ecology confronts a fundamental question: what is real? Is a biological cell, for example, a real unit (or "real whole"), or is it merely a simple sum of molecular reactions? And if a cell has a true, independent identity, then in what sense is a body composed of cells truly an independent identity? Or if the individual animal or plant is a real unit, then what standing has a larger unit like a family, a partnership, a species, or an ecosystem? Traditional science has always tended to assert that only the part is real and that understanding of the most basic element, gene, or molecule, is all that is necessary in order to understand all the higher levels as well.

But it seems almost self-evident (at least intuitively) that complex organizations like bodies have characteristics that do not arise simply as a result of the qualities and activities of their cells. An animal regulates its bodily state, moves about, thinks, chooses, feels, often sacrifices some needs in favor of others. These actions appear to be a product of the higher organization itself. The same insight seems to apply at each level; the higher level is, of course, influenced by and limited by its parts, but also seems to use them in a way that is unique to itself as a higher-order organization. Cells are not mere sums of molecules, but molecules bound into a very special kind of order. The science of ecology, with its interest in seeing how the many organisms work together in a functioning whole, has historically lent weight to this holistic alternative. Family groups, symbiotic partnerships, interacting species, and whole ecosystems all have been seen as possessing characteristics proper to them as higher-level organizations. In some important way, these larger units are also real.

The many ways of defining the "larger unit" continue to spark energetic wrangles among ecologists, which we will briefly consider later in this chapter. Yet, taking a step back, it is worth noting what an important door is opened, once wholes are readmitted to serious consideration and debate. Through this door one can glimpse a possibility quite missing from traditional science, yet quite familiar. It is a way of regarding nature that has moved the human imagination virtually forever: the harmony, beauty, and oneness of nature: the living world as more than a random assemblage of things, but a vast, intricately ordered unity.

❍ ❍ ❍

But what is this "oneness" of nature? How can it be understood not just religiously or aesthetically, but rationally? Is there a form of reason that can not merely take nature apart, but also put it together? This is vitally important, because humans experience things as wholes. A look at a landscape reveals some kind of essential unity that is not explained by an inventory of parts. Now this is a feeling, or an assumption, and no proof. Yet the experience is so real that for the reasoning mind to absolutely deny it is to work a kind of violence upon the integrity of one's experience.

Perhaps the most basic approach to recovering the unity of nature has been to compare this mysterious, larger unit—the ecosystem, the forest, the prairie—with a familiar smaller unit—the single body of an individual animal. The passage quoted at the beginning of this chapter about the "celestial orb" of the earth does it. So does this excerpt from *Sand County Almanac* (1949) by the pioneering ecologist Aldo Leopold:

> The most important characteristic of an organism is that capacity for internal self-renewal known as health. There are two organisms whose processes of self-renewal have been subjected to human interference and control. One of these is man himself (medicine and public health). The other is land (agriculture and conservation). The effort to control the health of land has not been very successful. It is now generally understood that when soil loses fertility, or washes away faster than it forms, and when water systems exhibit abnormal floods and shortages, the land is sick. Other derangements are known as facts, but are not yet thought of as symptoms of land sickness. . . . The status of thought on these ailments of the land is reflected in the fact that our treatments for them are still prevailingly local. . . . Many conservation treatments are obviously superficial. Flood-control dams have no relation to the cause of floods. Check dams and terraces do not touch the cause of erosion. Refuges and hatcheries to maintain the supply of game and fish do not explain why the supply fails to maintain itself. In general, the trend of the evidence indicates that in land, just as in the human body, the symptoms may lie in one organ and the cause in another. The practices we now call conservation are, to a large extent, local alleviations of biotic pain. They are necessary, but they must not be confused with cures. The art of land doctoring is being practiced with vigor, but the science of land health is yet to be born.

> A science of land health needs, first of all . . . a picture of how healthy land maintains itself as an organism.[8]

Leopold makes two points here which are absolutely fundamental to holistic ecological thinking. One is the ecological trait of looking at land problems (and, by extension, other problems) as a totality. The second is the metaphor of the single organism, which expresses that totality. Both points demand further attention.

A Holistic Approach to Problems: Net Cost Accounting

Behind Leopold's criticism of merely local solutions to environmental problems lie the same observations made by President Nixon's preface (quoted on page 14) and Commander Borman's meditation (quoted on page 12), in their different ways. Nature is no respecter of human boundaries and divisions, either intellectual or political. Its processes are so integrated and interlocked that to disturb one part is to disturb the whole. This recognition demands ecological problem solving that crosses state boundaries at will, and makes national borders mere imaginary lines. The Tennessee Valley Authority and its conservation goals were controversial in part for exactly this reason. The river valley the TVA administered was a natural unit that violated political definitions and delineations. The current acid rain problem similarly ignores the U.S.-Canada border. Both conservation in a multistate river basin, and pollution that drifts hundreds or thousands of miles, demand recognition of the unity of the earth. They make political distinctions seem artificial. But such recognition is an inherently radical idea, a biological version of "one-world-ism," usually a hated heresy in American politics.

Holistic thinking is what the Nixon preface calls for when it advocates "full accounting for the social costs of environmental pollution." Variously called "net cost accounting" or "whole systems analysis," this approach refuses to limit its vision to the linear, cause-and-effect simplicity of typical problem solving. The easiest and cheapest solution to a given problem is not necessarily the best, when unforeseen repercussions cause expensive problems elsewhere. It calls for a breadth of thinking that goes beyond artificial limits.

It is easy and cheap to produce a chemical and dump the side-product into the stream. But only if the problem/solution circle is drawn very narrowly—mainly around the company bank account. What happens to the effluents dumped into the stream? They may kill off the fish and aquatic life, spoiling the river's beauty—a loss in itself, and (if economic reasoning is required) probably a cause of lessened tourism and recreational income. Further down the line, the pollution will often cause health problems, which inevitably will cause lost days, months and years of labor from uncounted individuals. Society not only loses the productive fruits of their labor, but incurs an absolute drain on its financial resources in the cost of health care for the affected individuals. The result is a greater proportion of social resources expended in health care instead of in production of goods and services. And the net difference is likely to be considerable.

Researchers have, for example, established a positive correlation between air pollution and lung cancer, other respiratory disease, and mortality rate in general. Lave and Seskin reported a tenfold difference in death rates due to lung cancer between rural and urban English citizens. A similar study in the U.S. found that males who were lifetime city residents were twice as likely to die of lung cancer as their rural counterparts. (Both studies were adjusted for effects of tobacco smoking.[9]) The real cost of these losses in life is literally incalculable.

Can America afford pollution clean-up? Perhaps not if only the factory and its ledger sheet are looked at. But if the net cost of the entire situation is factored in, then even in the hardest-headed financial terms, the answer is undoubtedly "yes." To people used to exercising their "freedom" by ignoring the larger results, this too is a radical idea. It brackets private action within a very broad concept of responsibility.

But in fact, not only do the wider *effects* of pollution cost society money; the *creation* of pollution also represents a waste of money and resources. According to M.G. Royston of the International Management Institute in Geneva, "pollution is a sign of wasteful inefficiency and represents potentially valuable resources in the wrong place." In the early seventies, when most Western countries were putting about 1 percent of their gross national product into pollution control, Japan was investing about 6 percent. The result was not only a cleaner environment but also more efficient and profitable industries.[10] In the U.S., the 3M Company was applying similar techniques, and with the same results. It replaced add-on "end-of-pipe" pollution control with "product reformulation, process modification, equipment redesign, [and] waste recycle or reuse."[11] A wider vision of causes and effects produced an improvement in the whole social and economic performance.

Another example of net-cost or holistic accounting is offered in the book *Ecotopia Emerging*. This novel by Ernest Callenbach, like its companion volume *Ecotopia,* offers a remarkable view of the world envisioned by ecologism. Callenbach's Ecotopians are adept at whole-systems or net-cost analysis. When they look at the practices of twentieth-century America, they see countless instances where narrowly conceived public policy mistakenly throws systems out of kilter. As usual in utopian fiction, the purpose of imagining a new world is implicitly to criticize the old. Regarding that typically American form of politics, dam-building, an Ecotopian observes:

> The bioeconomic pattern is this: dams are built at costs so high that resulting irrigation water "could not" be sold at its true costs; hence it is heavily subsidized, normally by selling it at one-third or even one-tenth of its real-world costs. . . . The over-all effect of the dams has thus been to heavily subsidize farmers in irrigated districts, at thousands of dollars per acre, enabling them to out-compete farmers in rainy districts. The result is that, for instance, almost all the broccoli eaten in Massachusetts is grown in California and shipped east at heavy costs in oil for transportation, although it is in reality considerably cheaper to grow broccoli in Massachusetts during much of the year. Thus dam policy has substantially *diminished* the over-all efficiency of the national food-production system, causing a rise in the real total costs of food—though some of these costs are rendered "invisible" by being paid in taxes rather than at the supermarket.[12]

Hidden costs in agriculture, as elsewhere, are a widespread problem. As Eugene Odum remarks, "The public, and many professional specialists as well, have been misled by incomplete agricultural bookkeeping."[13] The ecological thinker takes the entire system into account, not just the farmer's immediate costs or the consumer's price tag at the grocery; any other approach seems simple self-deception.

Looking at almost any problem ecologically—in its larger context—leads to novel ideas of cost, effectiveness, and responsibility. Consider Ivan Illich's surprising evaluation of the true costs of private automobile ownership. In 1974, he declared that "the typical American male devotes more than 1600 hours a year to his car." This comes to four of every sixteen waking hours—sitting in it, parking it, searching for it, and earning money to buy its fuel, pay its taxes and monthly payments, repair it, and so forth.

> The model American puts in 1600 hours to get 7500 miles: less than 5 miles per hour. In countries deprived of a transportation industry, people manage to do the same, walking wherever they want to go, and they allocate only 3 to 8 percent of their society's time budget to traffic instead of 28 percent.[14]

The true radicalism of the holistic approach to problems is perhaps sufficiently visible in the examples given above. Whole-systems analysis is the "basic reform in the way our society looks at problems and makes decisions" advocated by the President's Environmental Council Report. If applied to other social problems, the results would be as surprising and revolutionary as those depicted in Callenbach's ecological utopia.

The Superorganism

The second point to be drawn from Aldo Leopold's meditations about the health of the land concerns the kind of language and metaphor he chooses to convey his sense of nature. It is the language of the organism—nature as a single body which regulates and heals itself just as a single animal does. This metaphor expresses vividly the holistic sense of unity and interconnectedness. It is language which produces surprisingly important effects in the way people imagine nature, the way they picture the world.

Frederic Clements, a founding father of ecology in America, based his science on this same analogy. He was a midwesterner, and was especially concerned with understanding that most American of biomes, the prairie. Clements felt that the only way to understand the prairie was to see it whole. There were intricate interactions between tall and medium and short grasses, between early and late bloomers, between quick growers and long stayers, between plant and weather and season and fire—and it was these interactions, together, which resulted in the almost indestructible grasslands that once stretched across half the continent. A given stretch of prairie was not made up of merely chance accumulations of individual plants; each species was an element of a closely integrated whole. Each species had coevolved over long eons to play a particular role in the grassland ecology. Clements found that if disturbed, as by plowing or fire or other disaster, the prairie would renew itself by predictable stages of growth. And in time it would regain its optimum state: the mixture of leafy plants, grasses, animals, and accumulated matter that made up the "climax formation"—what Clements regarded as the ultimate and permanent stage.

Understanding particular individuals or species alone seemed to give no clue about these dramatic facts of prairie self-renewal, interdependence, and longevity. Starting in 1904, therefore, Clements defined the prairie as a *single animal,* a "complex organism" which grew as an animal grows towards a last or climax phase, its "maturity." The prairie climax was "the adult organism."[15] Each successive stage amounted to "a superior life form," as the grassland became more complex, more stable, and nearer its final form. In the end, the prairie grew into a wholly stable association. "Such a climax is permanent because of its entire harmony with a stable habitat."[16] Thus when the later ecologist Aldo Leopold referred to the "health" of the land, he was employing less metaphor than one might think. Ecologists tended to mean it literally, up to and beyond the time of *Sand County Almanac:* the land, the ecosystem, could be literally regarded as a single, complex, living creature.

The science practiced by Clements and his followers in ecology described nature in terms of near perfection. Notice the permanence and "entire harmony" of Clements' description. A later ecologist (in the 1940s) saw the prairie as "approaching the eternal" in its self-integration and unity.[17] The prairie and other biomes studied by ecologists show the unexpected ability of plants and animals to coexist, to mesh, to find marvelous ways of adapting to one another and to the entirety of the physical environment. Ecology thus gives the lie to bloodthirsty or pessimistic interpretations of life on earth, such as the popular idea of Darwinism. Without mentioning it overtly, ecology declares the goodness, the beauty, and above all the *oneness* of nature. For those with eyes to see, as Clements described it, nature's apparent diversity and competitive chaos were merely local features. The true view of the matter—the perspective raised high enough to view the whole—revealed a profound harmony binding all into one living being.

The "complex organism" is a somewhat strange, but powerful concept. Since we are accustomed to thinking of individuals as the basic unit, it is imaginatively stimulating (perhaps shocking, perhaps liberating) to consider that individuals may themselves be tied together in some intricate way that is not obvious, but very real.

At about the same time that Clements was working, the entomologist William Morton Wheeler applied the concept of the complex organism to ants. His studies had become world famous for the depth and detail of the information they contained about these fascinating insects. Wheeler proposed to regard the ant *colony* as the real, single organism, and the ant as merely a cell within it. "An organism," according to his definition, "is a complex, definitely coordinated and therefore individualized system of activities."[18] These activities, essential to any organism, are primarily nutrition, reproduction, and protection. Only in the colony as a whole can the ant carry out these basic biological functions. Some ants can only forage; some only reproduce; some only defend. Only together are all three accomplished. Only together within the colony, therefore, are ants a true organism.

Recently, the influential biologist Lynn Margulis has followed up Wheeler's guess that perhaps big animals like dogs or humans were "essentially the same" as a colony.[19] Margulis hypothesizes that the first multicelled animals were in reality a sort of superorganism—a colony of simpler one-celled prokaryotes linked together into the first eukaryotes. The bodies of higher animals, from this point of view, are a sort of highly specialized colony of one-celled animals. Somehow, a higher unity pervades the organization; but ultimately they are still accumulations of one-celled animals.

How far could such an idea be pushed? Does nature present "real wholes" larger, even, than a prairie or forest ecosystem?

The atmosphere biologist James E. Lovelock thinks the answer is "yes." He thinks the whole earth might be integrated in a way that allows it to do just what Wheeler's ants do, or what any animal must do: live, breathe, maintain its parts and its vital functions such as temperature and chemical balances (though not, apparently, reproduce). Lovelock calls this huge organism "Gaia." And he offers persuasive (though still incomplete) evidence that in crucial ways such as maintenance of atmospheric carbon dioxide and oxygen, the entire planet may be integrated into a single, self-regulated system.[20] We will look at the details of his idea in a later chapter. It is the grandest application yet of the "compound animal" idea, offered as a serious scientific explanation for the systems that regulate planetary conditions.

The name for such a compound animal is the "superorganism." Clements' prairie climax is a superorganism: an animal or being made up of many individuals. Typically a superorganism is more spread out in space than most organisms we are familiar with, and hence it is easier to miss seeing. But it is, to these scientists, real nonetheless. The concept has been applied to many levels of organization: individual animal, ant colony, ecosystems like the prairie, human society, and beyond. The ecosystem itself is perhaps the most influential and important superorganism, since it provides the basis for ecologism and has been studied in detail. But, whether in cells combined to make an animal, animals and plants combined to make an ecosystem, or all together to make an ecosphere, the basic ecologistic leap is contained in the idea of the superorganism. This idea expresses forcefully the ecologistic concept that we overlook essential features of biological life if we focus on preconceived notions of identity. Life is capable of organizing itself almost limitlessly, from the cell to the organism to the superorganism. To draw an arbitrary line of separation between individuals or levels of organization is to miss their essential connection. Needless to say, this holistic view reverses the direction of typical scientific thinking, and opens up suggestive vistas of connectedness and relationship underlying the appearance of separation.

In effect, Lovelock has taken the green earth floating in the void seen by Borman, given her the name "Gaia," and declared that precisely what the earthrise picture *feels like,* it really means, in cold fact. The planet is alive. We are part of its life.

○ ○ ○

If nature in its forests and prairies, its animal species and individuals, and even its totality, may be in reality one vast breathing organism, then it is hard not to view it with an almost religious awe. In fact the idea of the world-as-organism is as old as the West—see Plato's *Timaeus*[21]—or even as old as humanity—as archaic worship of the "Earth-Mother" suggests. No history of "nature" in human thought could miss the central fact of the emotion of reverence and the impulse of worship which it awakens. What is new is that this fundamental impulse is represented through the agency of science and meets the criteria of rational, empirical, and determinedly nonreligious science. Nature-as-superorganism offers the pattern for a wholly different and more integrated style of thinking than has been available to the rational Western intellect. It leads the mind past surface details and scattered appearances, to look for the unity which enables them all to function as a whole. It includes the human species in the compass of its connectedness. Because of this, in the words of Gregory Bateson, it rediscovers the "sacred unity of the biosphere."[22]

The holism of ecological thinking thus leads *through* science to religious, ethical, moral, and imaginative participation in nature. It sees a fundamental unity out of which all these different kinds of meaning and response may independently and collectively arise. For even the human race, with its mind, is a part of this world-body. Its science has begun to describe the intricate functioning and interrelation of the living world, in ways which seem to give a basis for other measures of awe and delight.

Contemporary Science: Exceptions to Holism

Such a radical way of looking at nature has not gone unchallenged. Though the field of biology has typically produced a regular but small supply of renegades insisting on some form of holistic thinking to supplement the usual analytic approach, in modern times only ecology has been so thoroughly under their sway. The "organismic analogy" has been a decisive influence on ecological research and theory for many decades, from Clements to the present. In opposition to this, a countermovement has offered a more standard scientific explanation for ecosystem phenomena, but without much success until relatively recently. In the 1920s, H. A. Gleason proposed an "individualistic" interpretation:

> Far from being an organism, an association is merely the fortuitous juxtaposition of plants. What plants? Those that can live together under the physical environment and under their interlocking spheres of influence and which are already located within migrating distance.[23]

At its core, this is the issue of what kind of science biology ought to be—a perennial and still quite vigorous question. The example of classical physics continues to provide one model of a correct science: it is atomist, mechanistic, and reductionistic. It interprets all large-scale effects as simply the product of many smaller motions, just as the behavior of a quantity of gas is merely the sum of the motions of its molecules. The holistic approach provides another model: impressed with the self-integrated functioning of such wholes as the organism and the ecosystem, it searches for the rules of organization that make such surprising behavior possible. Does biology need to be "a science in its own right"[24]—or in Ernst Nagel's phrase, more than "simply a chapter of physics"?[25]

A continuing European tradition has more strongly held to the holistic side during much of this century.[26] In the English-speaking world, ecology is the only branch of science that has regularly employed a holistic approach. A holistic style of ecology enjoyed wide acceptance through the 1960s and '70s, for instance in the dominating standard textbook of Eugene Odum (in three editions from 1953-71). While Odum did not champion the Clementsian superorganism, he nonetheless depicted natural systems as very real, integrated wholes that had their own organization and self-regulation. "Community ecology" in particular, which deals with these large-scale interactions of many individuals, is "chronically" the scene of "tumultuous" disagreement, precisely because it focuses this basic question of how biology ought to be done.[27]

During the last ten years especially, a reaction along Gleason's lines against the tradition of ecological holism has gained strength. Understand the lives and deaths of individual plants and animals, say these reductionists, and you will understand the ecosystem. By extension, other wholes are also "reduced" to actions of their simple constituents: biochemistry will explain the body and mind; genes, not whole organisms, are the true locus of the crucial cause and effect of natural selection; and so on. To illustrate the strength which these scientists have achieved since the middle 1970s, consider the important general ecology textbook of Begon, Harper, and Townsend (1986), which issues this caution to its students:

> In Chapter 16 we cast aside the view that the community could be regarded as a sort of super-organism (the view of Clements). . . . when we start asking questions about how it comes about that these community processes occur the way they do, we are forced back onto the behavior of individual organisms for our explanations. It is they that are the material on which evolution works, and it is they, through their main effects and interactions together, that account for the activities of the community as a whole. We must not forget this.[28]

But despite such strong, partisan statements, the debate over the degree of holism needed in biology remains open. The evidence that living beings are organized on higher levels than the individual atom or individual part continues to impress some scientists with the need for an approach that does not stop with the kind of interactions found in physics, but goes on to embrace the special cases of biology. As quoted in Robert P. McIntosh's excellent review of recent ecological issues, the eminent scientist L.B. Slobodkin wrote in 1965.

> The normal criteria of scientific quality which we use as biologists, are not the same as those of the physicist and mathematician. . . . Empirical sciences must develop their own standards of quality and cannot take refuge from the necessity of thought in the shadow of Newton or Euclid.[29]

McIntosh could characterize many of the essential ecological (and ecologistic) concepts—such as "climax, stability, and equilibrium"—as "much in dispute" in 1980, mostly because of this underlying question of how to define a scientific technique appropriate to biology.

Others have continued the debate up to the present. William C. Wimsatt (of the Department of Philosophy, Committee on Evolutionary Biology, and Committee on Conceptual Foundations of Science, University of Chicago) has written extensively to show the limits of the reductionist approach, as for example whether genes or organisms are the units of Darwinian selection. In general, he argues, the reduction of natural effects to effects of individual parts is not a description of nature, but a heuristic device—i.e. it may be useful, but is not necessarily true. Further, he notes, "the problem solving heuristics used by reductionistically inclined scientists result in systematic distortions biasing the case."[30] A reductive approach will tend to screen out all effects that cannot be accounted for on the lower level, thus creating a self-perpetuating distortion.

In his depiction of reductionism as merely one scientific tool among many, Wimsatt is joined by Richard Levins and Richard Lewontin, both of Harvard University. They point out the necessity of getting beyond "vulgar reductionism" into a kind of open-minded pluralism, on the grounds that a simplifying theory of nature will always fail to do justice to a wild, fecund, and surprising biological reality:

> . . . that nature is contradictory, that there is unity and interpenetration of the seemingly mutually exclusive, and that therefore the main issue for science is the study of that unity and contradiction rather than their separation either to reject one or to assign relative importance.[31]

The openness of the ongoing question about the holism or reductionism of biology is not restricted to academic philosophers. While one recent textbook, as we have seen, takes a very hard stand against holism, others are less sure. The introductory text *Principles of Ecology,* by Putman and Wratten, for example (which was published in 1984 and reviewed glowingly in the professional journal *Ecology*[32]) talks quite openly about "whole systems," one of the concepts avoided by the Begon textbook. It does not dismiss the idea that a real entity may exist at higher levels than the simple or primary ones envisioned by reductionists.

An example of this openness to nonreductionist solutions concerns a very fundamental ecological question: how the numbers of individuals are distributed in ecological communities. This has been a central issue from the time of Clements, who held that an ecosystem contains the organisms it does, in the amounts it does, because these organisms and amounts are somehow selected by the ecosystem itself, in a process which results in optimum functioning of the whole system.

Putman and Wratten first offer the reductive explanation of R.M. May, whose mathematical model suggested that the often observed, strikingly patterned distribution was a merely mathematical and individualistic effect, which would "not necessarily reflect any intrinsic characteristic of the community itself."[33] May's explanation is a typical reductionist/individualist alternative to the holistic approach. The patterns observed are apparent, not real. The patterns cannot cause anything to happen, because they only exist as the result of lower-level events. An ecosystem is only apparently a system at all (Begon, in fact, avoids the term altogether). The seeming order and organization of the ecosystem are epiphenomena produced by true cause and effect at lower levels.

But Putman and Wratten—and this is the important point—offer another side, as well. When a researcher examined *real* communities (in contrast to May's mathematically modelled one), he found that the patterns there "are indeed a function of specific characteristics of the community."[34]

It may be worth noting the general structure of this episode: a mathematical model fails to predict the surprising behavior of a real system—evidently a frequent occurrence.[35] What the German scientist Adolf Portmann criticized as "the physicists' dream of a mathematically ordered world"[36] is unable to capture the real behavior of organic systems—what a writer in the journal *Science* has called "the essentially more complex, three-dimensional world of whole organisms and the communities in which they live."[37]

The holistic side of ecology is controversial, and is opposed by a mechanistic and reductionistic approach that describes living phenomena in the terms of Euclid and Newton. But the complexity of organisms continues to baffle simplifying explanations. Biologist and historian Stephen J. Gould (writing specifically about the genetic determinism aspect of this issue) concludes that it is the "bad habits of Western scientific thought . . . atomism, reductionism, and determinism" that mislead biologists into ignoring the holistic side. While the analytic style works well for the inanimate world, it does not provide a complete method for study of the living world.

> Organisms are much more than amalgamations of genes. They have a history that matters. Their parts interact in complex ways. . . . Molecules . . . are poor analogues for genes and bodies.[38]

It is perhaps the special burden of ecology, as a science, to maintain some sense of the wholeness of the organisms and communities which it studies. Biology may not be, after all, physics. Living organisms may be more, and other, than what nonliving assemblages of matter are. They make wholes which have characteristics of their own, not proper to any lower level of analysis. These wholes are evident to us experientially; and our experience does not *have* to be falsified by our formal knowledge—if it is possible to integrate this holistic side into it. As Gould finally puts it, "My intuition of wholeness probably reflects a biological truth."[39]

It is this connection of experience with knowledge, of intuition with scientific truth, that generates the galvanic spark energizing the ecologistic worldview. As I commented earlier—and as I will reiterate in chapters to come—this worldview does not keep itself in perfect alignment with the precise but shifting discussions of science. Ecologism tends to ignore some of the inconvenient data and theory. But at the same time, it draws strength and encouragement from the real commitment of scientific ecology to something very special among the sciences: a certain quality of seeing the whole.

The Major Myth

For many decades, Clements' superorganism was the emphatic form of holism that most influenced ecology. And its influence has spread out, touching our very way of thinking about the world. The superorganism is what I call a "major myth" of the ecological worldview. By myth I do not mean "falsehood." A myth is an image or story that encapsulates some primal truth about the world. A myth is an idea with a power beyond the rational and limited impact of the literal. A myth is used as a pattern for understanding. And it is used as a standard for behavior.

For example, the Pilgrim Fathers are a familiar American myth. They came to the New World fleeing persecution and seeking liberty through great hardship. Many others came, before and after, for many reasons. Why single out the Pilgrims? Because they are our myth: the fact which embodies something deeply exemplary, something which explains who we are and sets an ideal of behavior.

In a similar way, the worldview of ecologism lays hold of certain representative pictures or examples. The superorganism expresses the idea of holism in a compact and concrete form. It is a reference point for the imagination, summarizing the fundamental truth of ecology: that living things are so deeply interrelated, so profoundly dependent upon each other, that their lives are in effect one. This interconnection is a scientific fact, undisputed by either side of the debate.

As we have seen, scientists and philosophers have argued vociferously among themselves about the literal reality of many forms of the superorganism. Clements' theory is decidedly passé as straight science; but descendants of it, in the various styles of holistic logic described earlier in this chapter, still play a role in ecology. But regardless of the superorganism's somewhat demoted status, it plays an important role as a major myth of ecologism.

As ecological thinking spreads out in our culture, and sinks more and more deeply into our habits of mind, the superorganism appears in unexpected places, repeatedly and insistently. The popular essays of Lewis Thomas, for instance, have given it wide currency. In his best-selling book *The Lives of a Cell,* he returns again and again to the central image of the superorganism. What is the earth like? "It is *most* like a single cell."[40]

Thomas tells story after story illustrating the myth—the definitive pattern—of superorganism. The familiar termites, ants, and bees appear early (these are favorite eco-stories, retold in many places as illustrations of the surprising unity of life). Soon, however, Thomas offers this unusual little gem, which I reproduce for its vividness:

> The phenomenon of separate animals joining up to form an organism is not unique in insects. Slime-mold cells do it all the time, of course, in each life cycle. At first they are single amebocytes swimming around, eating bacteria, aloof from each other, untouching, voting straight Republican. Then, a bell sounds, and acrasin is released by special cells toward which the others converge in stellate ranks, touch, fuse together, and construct the slug, solid as a trout. A splendid stalk is raised, with a fruiting body on top, and out of this comes the next generation of amebocytes, ready to swim across the same moist ground, solitary and ambitious.[41]

"Solid as a trout": Thomas wishes to get across that, for him, the superorganism is no mere metaphor. It is a biological fact, a reality we have too often missed in our limited view.

Once this pattern is set, Thomas moves on to find the organic unity in all manner of supposedly separate phenomena. Between the extremes of the micro-superorganism and the world-superorganism stands humankind. Not only is the human body a colony of quasi-independent cells; each of us is also extraordinarily closely joined into larger bodies of social and mental connectedness. "Although," writes Thomas, "we are by all odds the most social of all social animals—more interdependent, more attached to each other, more inseparable in our behavior than bees—we do not often feel our conjoined intelligence." Humans are "compulsive" communicators who share an intense connectedness through language. "The human brain is the most public organ on the face of the earth, open to everything, sending out messages to everything. . . . We pass thoughts around, from mind to mind, so compulsively and with such speed that the brains of mankind often appear, functionally, to be undergoing fusion." "Perhaps," he speculates, "we are linked in circuits for the storage, processing, and retrieval of information, since this appears to be the most basic and universal of all human enterprises."

> This is, when you think about it, really amazing. The whole dear notion of one's own Self—marvelous old free-willed, free-enterprising, autonomous, independent, isolated island of a Self—is a myth.[42]

(Thomas uses the word "myth" here in its popular sense, i.e. "falsehood.") It takes only a little imagination to see humans in organizations and societies, or even as a whole species, as essentially linked, just as the ants or termites are linked. We are not characteristically "human" at all, except as participants in our language, family, work, and other social groups. To Thomas, our cultural training and indoctrination as rugged individuals is hardly more than whistling in the dark; so powerful is the similarity of all life, its essential oneness, cooperativeness, and readiness to join, that only by considerable effort do life forms keep themselves distinct at all.

❍ ❍ ❍

The superorganism takes perhaps its most extreme form, however, in the writings of the French Catholic mystic, Pierre Teilhard de Chardin. The future, says Teilhard, is moving toward a convergence of biological and mental life. Eventually, at what he melodramatically calls the "Omega Point" of history, the minds of planet earth will reach a critical mass and, *voilà,* fuse themselves together. (How strong a resemblance to Thomas' parable of the slime mold!) This is the direction of evolution. Teilhard's name for the fused earthly mind is the "noosphere": "the pan-terrestrial organism in which, by compression and arrangement of the thinking particles, a resurgence of evolution . . . is striving to carry the stuff of the universe towards the higher conditions of a planetary super-reflection." It is the superorganism writ large, and carried forward to a religious apocalypse.

Teilhard's popularity rose with the rise of ecologism; the two express a remarkably similar ethos of connectedness. And both use science as the springboard to prove it. Teilhard's works are well salted with technical words from physics and biology, even with scientific-looking diagrams and axial coordinates. His essay "My Fundamental Vision" (quoted above) begins with the insistence that "the totality of life" be examined "scientifically."[43] Teilhard's use of science is purely poetic, however. To follow his logic is to accept analogy in place of proof, and to share his vision is to partake of a *faith* in humanity's future, not an analysis of its present. Worldviews are typically analogical, too, as I have already said. But there is a crucial difference in the amount of faith they demand. A worldview describes the visible and experienced world in a way that makes immediate sense. It convinces almost before the mind has had a chance to analyze or reflect. Its analogies are felt to be persuasive because, like any good explanation, they make sense of so many things, bind them into a unified pattern. They express what has already been experienced. In contrast, Teilhard's flight into the future shares only a little of this quality. For most readers, it remains more of a vision and an exciting conjecture than a literal possibility. In addition, Teilhard's vision also dramatically contrasts with ecologistic thinking in that it is purely anthropocentric, a point made forcefully by Deep Ecologists George Sessions and Bill Devall.[44]

That Teilhard's vision has gained such wide notice and appreciation, however, may be a telling indication that the instinctive individualism of the Western mind is undergoing a deep change—that wholeness and connectedness are becoming increasingly important values. For these values, the superorganism is a splendid expression.

Practical Holism

If Teilhard expresses the mystic edge of holistic thinking, there are plenty of examples to show that a holistic approach can be practical and down to earth. In the category of practical results, one need only look at the geodesic dome. This is the first new form of architectural span since the pointed arch, and it is a creation of the first magnitude. Its inventor, the late Buckminster Fuller, became a world celebrity. When he spoke or wrote, his message was usually the same: strength lies in the unity of parts. This is the principle of the geodesic arch, in which relatively thin and weak members are joined in a star-like, angular pattern that creates the lightest, strongest span known to engineering. Fuller's word for the strength that is more than a mere sum of the parts is "synergy"—a word used years before by William Morton Wheeler, to describe the curious joint intelligence of many tiny mindless ants. Throughout his career, Fuller made a point of trying to bring holistic thinking to bear on problems, trying to work from the big picture, where everything is connected, down to the specific problem to be solved.

Holistic thinking is familiar on a practical level to millions of ordinary citizens in the form of holistic health. While burdened with an unfortunate amount of quackery and faddism, holistic health has nevertheless made some solid contributions to public and professional health care. It has fostered an awareness that a person is not just an assemblage of parts, but a self-integrated system whose normal state is balance—"health."

The holistic approach is most appropriate where there are many causes or parts interacting. Complex interactions like the weather, the economy, the functioning body, or the ecosystem probably cannot be fully understood by traditional analysis, because each part influences the behavior of so many other parts, each cause might have many effects, and the higher levels of organization in a complex system may not be obvious to an observer at all. The techniques for understanding such interactions are very new, and still far from perfected. Computer simulations, to date, tend to be too simple to be very reliable; though as more detailed models are constructed, and as more powerful computers are brought to bear, great improvement seems possible. However, these can be criticized as linear, not holistic; it remains to be seen whether they can truly model complex, hierarchically overlapped and layered real systems.

In short, the complexity of the real world demands holistic approaches as a complement to the more familiar analytic ones. But we still have much to learn about holistic problem solving.

Conclusion: The Heart of Ecologism

The holistic way of thinking lies at the core of the ecological worldview. It gives rise to the most fundamental differences from traditional ways of thinking, and leads to new insights and new practices on almost every level of human endeavor, from the purely spiritual to the most practical.

The ecosystem provides the example, the model, for the holistic view of nature and the world. As pioneer scientists like Frederic Clements began to uncover the unexpected integration of natural systems during the first years of this century, it became evident that much was missing from the accepted views and techniques of science. Their search for explanatory ecological models led them to interpret nature in terms of the organism, instead of interpreting it in terms of the atom; the whole, instead of the reductive part.

The expression of holism in the image of the superorganism provides an imaginative handle on a difficult concept—difficult, at least, for minds trained to take things apart. The idea of a vast and mysterious unity feeds the creative and emotional side of the mind. It is possible to make poetry of this, and ethics, and religious insight. It grounds the private experience in a rich field of shared and intersecting realities—one's participation in mental, bodily, and ecological wholes.

The superorganism is only one of the important myths that carries the ecologistic message of integration and connectedness. It is the broadest and most encompassing, as illustrated by the startling "Gaia," the planetary being in and through which all of us live. As we shall see, other myths of ecologism present the idea of connectedness on levels closer to home.

CHAPTER III

Balance

"There was once a town in the heart of America where all life seemed to live in harmony with its surroundings." So begins one of the most influential ecology books of the twentieth century: Rachel Carson's *Silent Spring*. It sounds like the place you and I would like to live, if only our livelihoods and responsibilities would let us. A human settlement, to be sure. But enclosed in the natural harmony, not out of touch with the forests, the animal lives, the seasons.

And enduring. "So it had been from the days many years ago. . ."[1] There is no sense of pell-mell change in the fable with which Carson starts her book. The town is weathered and familiar. Its population doesn't threaten to "explode." Just as the meadows and forests around it can be counted on to change shape only very slowly, and to be there for the returning eye many years hence, so one feels this town will be the same town always.

This appealing picture speaks to one of the most powerful of human longings: the desire to be rooted in nature. It is true that feelings about nature are a cultural product—a learned behavior or set of attitudes. But it is also true that almost every era and society has found a way to express the harmony and permanence of nature. And frequently this enduring harmony is contrasted with the temporariness and artificiality of the city.

In the ecological worldview, both the beauty and the enduringness of nature are a product of one crucial attribute: its *balance*. For each fast-multiplying grazer or browser there is a predator; for every predator there are forces that keep its numbers in check. For every dying forest tree there will be one to replace it. For each molecule of water that flows down into the sea, there will be one lifted by the power of the sun and sent inland over the plains and forests, to begin the cycle again. The balance of nature holds all together and carries it into the future. And nothing in this balance can be wasted, because nothing that is truly natural can disrupt the processes of creation, dissolution, and recreation, the cycles of building up and tearing down.

This idea of balance is worth examining in greater detail, for it embodies one of nature's most important attributes. In the natural balance, the ecological worldview finds an answer to some of the more destructive values and beliefs of the West. And as the balance is worked out in both literal and metaphoric terms, it exerts a persuasive influence over minds which are ready to turn away from the one-sided, human-centered, technologically bewitched and polluted life of modern industrial culture.

The Living Balance

Without the benefit of detailed factual information, it would be easy to imagine nature to be unchanging in the simple sense: an established set of relations between so many animals, so many plants, and the four seasons. In the most traditional of Western views, the species were in fact seen as *absolutely* stable, the hills as "eternal," the seas as established by divine creation. The harmony of nature was just this perfection of dependability, rooted in the unchanging Creator himself.

The ecological idea of nature both repeats and refutes this picture. Nature is certainly dependable and life-sustaining. But its balance is hardly changeless. Conceptually we may see nature not as a single ordained thing, but as a dialectic of paired opposites: birth and death, predator and prey, photosynthesis and oxidation, winter and spring. Like a piston, each natural act has its opposite, each stroke its reciprocating return. Yet the reality of the natural system is more complex than this. It might be better represented by the familiar image of the "web of life." This at least captures the many-branched unity of living things.

But this expression also falls short. It is too static. For the natural "balance" is anything but stationary. Nature's balance is *alive.* Perhaps a better image would be that of a stream, which retains its form but is never twice the same. Carson states:

> The balance of nature is . . . a complex, precise, and highly integrated system of relationships between living things. . . . The balance of nature is not a *status quo;* it is fluid, ever shifting, in a constant state of adjustment. Man, too, is part of this balance.[2]

But at the same time, the net result of the natural equilibrium *is* a kind of changelessness. These two elements—change and stability—are woven into the ecological worldview, and provide it with important (though sometimes contradictory) images about how the world is.

○ ○ ○

If nature is a "web of life," it is one in four dimensions, with a time component allowing each and every relationship, each junction in the web, to flex and reposition itself continually. It is doubtful that this web could ever be the *same* web twice; its interrelations would always be similar yet unique. Most of them would be undergoing cyclical change, oscillating within certain limits in response to other parts of the web. The name for this balance is *dynamic equilibrium,* which captures in two words both the ceaseless flux of the ecosystem and its enduring stability.

The ubiquitous little recycled paper symbol expresses the changing/unchanging quality of ecological nature in a nutshell. This is virtually a diagrammatic form of the Earthrise picture; the self-enclosed unity and circularity is powerful in both. The fluid stream of change, the flexing web of life, the recycling symbol, the Earthrise picture—all these metaphors and images seek to communicate the same basic idea of a stable but always-adjusting nature. They all project a sense of shifting interrelation, joined into a powerful overall

plan or harmony. The recycling symbol in particular reminds the viewer that nothing in the natural balance ever exists in isolation: having entered the system, it must flow through many changes and touch many parts of the web. Nothing that is placed in a balance can avoid exerting a weight against something else, or avoid calling out a counterpressure somewhere else to compensate. Every cause in a self-enclosed, endlessly cycling system has many effects.

In a day when this aspect of nature was not widely known, Rachel Carson's task was to communicate the balance, interconnectedness, and harmony of nature in a way that emphasized these cycles. The danger of nonnatural substances like DDT could simply not be understood while thinking was still locked in a simplistic, cause-and-effect style. The significance of a pesticide did not end when the offending mite, fungus, or insect was killed; once introduced, the chemical could circulate through the system almost indefinitely, until it finally caused drastic harm in some other locale.

Probably no other book in the huge library of ecological works has had a greater impact than Carson's in disseminating this notion of cycles and circular connectedness. The frightening accumulation of DDT in lakes, the shells of birds' eggs, and even human tissue was only the most spectacular part of this story. Carson shows that the cycles of nature are omnipresent and essential. The soil, for example, is pictured "in a state of constant change, taking part in cycles that have no beginning and no end." These cyclical processes reach even to the most fundamental levels of life: the places where the food chains begin (in photosynthesis) and end (in cellular oxidation). In the latter, for example,

> the transformation of matter into energy in the cell is an ever-flowing process, one of nature's cycles of renewal, like a wheel endlessly turning. Grain by grain, molecule by molecule, carbohydrate fuel in the form of glucose is fed into this wheel . . . . When the turning wheel comes full cycle the fuel molecule has been stripped down to a form in which it is ready to combine with a new molecule coming in and to start the cycle anew.[3]

These "endlessly turning wheels of oxidation" are miniatures of the entire network of self-sustaining yet ever-changing natural processes.

The balance of nature is a complex idea. It is a balance that in cold fact includes all elements that touch the living world. It resists compartmentalization in the way a circle resists squaring: no amount of straight lines can really describe it. No sequence of simple "A plus B" effects can exhaust the interactions, subtle and gross, of the natural web. Hence it is not enough to "balance" the cities with a few so-called wildernesses. This is a static notion of balance. What must be preserved is the dynamic *process* of which "wilderness" is merely the result. Taking out the wolf, the grizzly, the coyote, and the puma, and sending overhead clouds of acid rain and pollution, strike at the cycle itself. No neatly drawn mix of urban and natural areas can be effective if it ignores the fact that these are interconnected. For, as is often forgotten, "Man too is part of this balance." If the cycles begin to be disrupted, whether by dams for flood control or poisons for predator control, the disruption may unexpectedly turn up in a cycle closer to home; maybe even in the very cells of our bodies, as happened with DDT.

This is why the recycling of waste materials is almost a religion with ecologically minded people. It represents not just a particular bottle or can, with its fraction-of-a-cent economic value. Recycling represents nature itself. In the balanced system of the forest, dead trees occur no faster than they can be broken down, used by myriad other living things, and replaced. This is a cyclical process perpetuating the rich diversity of life. In the unbalanced artificial economy of the city, waste materials accumulate much faster than they can be naturally broken down; this is a one-way process ending in an ugly uniformity of litter. Likewise many manufactured wastes cannot be recycled and are therefore not merely nonnatural but *antinatural:* they violate the very structure of nature. The controversy over creating and stockpiling quantities of nuclear waste is more than a simple disagreement over practicalities (where to store it, or how). It is a conflict of visions.

Stability and Diversity

The ecological vision of nature, and humankind in it, emphasizes the long-term stability of the biosphere. It sees the processes of creation and destruction as equally precious and equally necessary to life. What matters is fitting in to these processes. What is dangerous is introducing changes or stresses that cannot be accommodated by the plants, animals, and environments that give our species its breath, food, and life. Excess can be tolerated in such a system only within limits; too much radiation, too many people, too many wastes—any of these, in sufficient quantity, could disrupt and damage the natural system, jolt the natural equilibrium past its ability to adjust. Without these disruptions introduced by sometimes unnatural human acts, the natural equilibrium can be both stable and diverse on a vast and awesome scale; extending into every living cell on earth, creating an astounding diversity of niches and roles, and—according to the Gaia hypothesis— self-adjusted over the span of thousands and thousands of years.

But the result of disruption is, typically, reversion to a vastly simpler ecosystem (assuming something short of sterility). Pumping sewage into rivers, for instance, creates anaerobic pollution; the rivers become transformed into systems of lessened diversity. In place of the trout or catfish and many varieties of insect and plant and microscopic life that had evolved over millenia to share the river, the flood of nutrients (sewage) stimulates a sudden growth of microorganisms and algae, whose rapid life-processes soon deplete the river of its vital oxygen. What is left is typically a simpler system favoring only very few, and relatively simple, plants and animals—the smelly, algae-laden result most familiar near large cities.

The tendency of human civilization to result in reduced variety in nature is significant for many reasons. On a purely practical level, there is the sobering importance of the gene pool. Genetic information that passes permanently from the world of the living can never be recovered. And whatever might have been the future evolutionary role of a destroyed species is lost. In a terrible and ever-widening sense, the future is impoverished by the thoughtless destruction of even a minor plant or animal species. There is no knowing what powers may have lain locked in these genetic banks. Less important, though still worthy of notice, is the human use of genetic resources which is also foreclosed by such extinctions. Standing on the threshold of biotechnological engineering, humankind would be prudent to safeguard the stock of living things. For they cannot be replaced, and theirs is the stuff of which biochemical and medical advances will be made.

The wholesale disruption of habitats and concurrent extinction of organisms is not a small contemporary problem. All over the globe, vast changes are being worked. For the last ten years, tropical deforestation for timber, agriculture, beef ranching, and firewood has proceeded at about the current rate: 80,000 square miles per year (an area about the size of Kansas), according to biologist Peter Raven's 1987 keynote address to the American Association for the Advancement of Science. At that rate and assuming no increase, tropical forests will disappear in thirty years. These ecosystems are almost fabulously rich in species, many or even most as yet unrecorded. To give one example, the rain forests of tiny Costa Rica—disappearing like forests everywhere in the tropics—harbor more bird species than *all* of North America.[4] Perhaps one half of all the world's species inhabit these forests.[5]

But such devastation is not limited to faraway equatorial regions. Just within the last few years, a massive decline or dieback of temperate forest has been recognized. Across the whole of Europe, the combined effects of industrial civilization have created the "Waldsterben" that now affects over half of West Germany's forests, and significant fractions of other nations' forests. The dying forests "look as if they've been sprayed with some chemical warfare agent"; older trees are yellow, spindly, and stripped of foliage; young trees grow into oddly distorted and deformed shapes; huge numbers die. Naturally, when the trees go, everything else goes too: "It is as if the entire ecosystem has been poisoned." And a decade or two behind Europe, North America is showing clear signs of following. Severe dieback is now obvious in higher elevation coniferous forests, such as those on North Carolina's Mt. Mitchell. A recorded slowdown in tree growth mimics a similar slowdown several decades ago in European trees; the forest systems are under tremendous stress, and this growth decline will be followed by dieback.[6]

Under such conditions of almost universal assault on earthly ecosystems (especially the richly diverse tropical systems), it follows inevitably that literally uncountable numbers of species—insects, plants, birds, and mammals—will disappear from the planet as their habitats fall to chainsaw, bulldozer, agribusiness, and pollution. Peter Raven concurs with many other biologists in estimating that presently the earth is losing "perhaps a few species a day," and that the rate will accelerate as the forests shrink and eventually (apparently) all but disappear.[7] Such losses constitute an incalculable narrowing of our planet's and our own future.

These losses must concern any sane human for more than practical reasons. Not only the world but we ourselves are fundamentally diminished when any element, any other species, is removed from the complex balances of life. Ecological thinking wishes to place the human being *within* the natural world, not above it. This means accepting a role as one among a myriad species. Our place in this sense is defined by the surrounding community of all these other lives. *Deep Ecology* calls this idea "biocentric equality":

> that all things in the biosphere have an equal right to live and blossom and to reach their own individual forms of unfolding and self-realization. . . . that all organisms and entities in the ecosphere, as parts of an interrelated whole, are equal in intrinsic worth.[8]

Aldo Leopold, forty years earlier, had enunciated nearly the same view that *maintaining* and *valuing* the richness and diversity of the natural order were equivalent to truly "natural" behavior: "a thing is right when it tends to preserve the integrity, stability, and beauty of the biotic community. It is wrong when it tends otherwise."[9]

○ ○ ○

The connection drawn above between the stability of an ecosystem and its diversity has been an important one for ecology, and also for the ecological worldview. It seems to link together two basic and very emotion-laden values: the desire for some degree of permanence, and the delight in a various and richly diverse world. Stability and diversity (or "complexity") are essential ideas, because without some form of each, and some combination of both, the orderly world of our experience would obviously be impossible: either a wildly fluctuating chaos, or a sterile monotony.

The ecological idea of balance seeks to express how the two qualities interact to create the intricate and dependable world we inhabit. As I have pointed out, this balance is a complex one, expressed in the near-contradiction "dynamic equilibrium." The ancient idea of unchanging nature is set aside; in its place, an idea of a fluid but resilient natural order, capable of flexing yet retaining some kind of essential, underlying organization.

Beneath this dynamically stable flux, ecology recognizes the unidirectional change of evolution. No community is thought to be permanent or stable on the evolutionary time line; time brings new conditions and works irreversible changes on organisms and the communities they make up. But within the longer rhythms of evolutionary change, the interrelations of organisms may be organized into enduring patterns—perhaps even coevolved in ways that lead to large, well-organized, relatively stable ecosystems.

For many years, the accepted formulation was that natural diversity and complexity *caused* stability. The 1970 Environmental Council Report put it this way: "The more interdependencies in an ecosystem, the greater the chances that it will be able to compensate for changes imposed upon it."[10] The ecologist Robert May has complained,

> Many well-meaning environmentalists embraced this "law" as providing a "scientific" justification for preserving complexity and diversity (because it fosters stability, self-evidently A Good Thing). . . . Further investigation and thought has shown the truth to be more complicated and various.[11]

May here shows some understandable impatience; the scientist bridles at the use to which his science is put by outsiders. And this example is one often singled out as a particularly glaring abuse. Perhaps the nonspecialists can be forgiven, however, for the idea was widely held within the scientific community as well, from at least the time of its promotion by the eminent Charles Elton in 1958 until the mid-1970s.

So what *is* the connection between those two components of the natural balance, complexity and stability? Some recent formulations have decided the process usually works in reverse:

> A predictable ("stable") environment may permit a relatively complex and delicately balanced ecosystem to exist; an unpredictable ("unstable") environment is more likely to demand a structurally simple, robust ecosystem.[12]

This is the conclusion originally proposed as a result of another of Robert May's mathematical models, and is referred to as May's Paradox, because it contradicts the "intuitively sound" opposite notion—for which, confusingly, there is nevertheless *some* experimental evidence![13]

Evidently no single, axiomatic outcome is in the immediate offing. "Stability" may be variously defined (focusing on various technical qualities like "resilience," "persistence," "constancy," and so on); "complexity" too may be defined and quantified in several ways. Answers to how the two qualities relate are thus highly varied. For example, a very complex, well-established ecosystem, such as a forest in the climax phase, will show great resistance to perturbation, and be stable in that sense (bearing out the "complexity begets stability" idea). But that same forest will not snap back very quickly, or perhaps at all, once it *is* shaken. Yet a very simple system might show just the reverse behavior.[14]

The connection of complexity and stability, however, remains a potent idea for several reasons. Firstly, whatever the direction of cause and effect, linkage exists and is intellectually interesting and emotionally resonant. Even if expressed in the one-sided terms of May's Paradox—that complex ecosystems occur in stable environments, and so are unprepared for severe stresses—there is a powerful lesson. Humans who value the richness of nature had better take care not to overstress the (apparently) frail systems.

Secondly, there remain interesting ways in which there does seem to be a significant causal connection from complexity to stability. When "diversity" is defined in terms of "energy exchange pathways," then a system in which many species are available to consume and pass on the system's energy is likely to be highly stable ("resilient"). (Equally stable would be a low diversity of very generalized animals—but though theoretically possible, this situation is rare or nonexistent in nature.[15]) This connection of energy pathways to diversity and stability bears on the question of cybernetics and the ecosystem, which is examined in a later chapter.

In some ways, a stable environment allows a complex natural system to develop. In some other ways, a complex natural system may have certain qualities which give it advantages over simpler ones. The connection of stability and diversity remains a subject of considerable scientific attention, and resists simple formulation.[16]

○ ○ ○

Perhaps partly for this reason, the most recent and important formulation of ecologistic thinking, Sessions and Devall's *Deep Ecology* has tended to forego offering a scientific rationale for maintaining the diversity of the natural world. While diversity is a central value of this book, it is usually defended on philosophical rather than scientific grounds, despite some scientific statements, such as in Arne Naess' somewhat obscure comment that "From an ecological standpoint, complexity and symbiosis are conditions for maximizing diversity." The somewhat strained relation of *Deep Ecology* to ecologism's scientific materials is examined in detail in a later chapter, also.

Others in the ecological movement, however, are deeply aware of the scientific aspect of the contemporary assault on diversity, and are acting on this knowledge. A most interesting example is Gary Nabhan, who writes in *CoEvolution Quarterly* about his work "helping Indian farmers to locate and conserve native desert-adapted crop varieties, since it is projected that over half of the crop varieties utilized in the New World at the time of Columbus may have already been driven to extinction."[17] Nabhan points out how the diversity of traditional agriculture has helped maintain the diversity of the surrounding ecosystem. Some of the southwest Indian villages have been in continuous habitation for many hundreds of years; "this stability through time has helped harbor agricultural diversity." (Note that Nabhan states that stability begets diversity, not the reverse.)

Nabhan shows a fine sense of the many-sidedness of this question: the linkage of desert agriculture to its entire region ("the whole web of interactions"); the large threat of "Earth [losing] part of the diversity of its life-support system"; the advantages of diverse, locally native farming over an imposed, alien monoculture ("to fit the crop to the environment rather than trying to remake the environment to fit the crop"). Most importantly, Nabhan shows how these ecologically sound practices are grounded and held together by a deeply spiritual sense of place: how the "Hopi sense of spiritual propriety" has maintained a detailed attentiveness to place and plant. These are traditional communities "in which the way that one farms and concerns himself with wild resources has *everything to do* with the spiritual life of the community."[18] It is this spiritual element that is so fatally lacking in supposedly utilitarian Western land practices. Nabhan's writings and involvement in restoring native seedtypes demonstrate a thoroughly ecologistic blending of science, environment, culture, and spirituality.[19]

The Climax Myth

It might be truthfully said that all ecosystems represent the ecological concept of stability-in-change. But none does so as perfectly, or as memorably, as the climax formation does. Accordingly, this is the mythic image in which ecological thinkers, writers, and poets most frequently encapsulate the idea. It provides an apt illustration for the serene self-adjustment of nature, its beautiful diversity and interconnectedness, its processes of living change, and its permanence.

The climax formation is an enduring and stable mixture of plants and animals, described by Frederic Clements and his followers, as we have already seen, as the "perfect" life form for a given climate and locale. The climax phase is led up to by a series of stages known in general as "ecological succession." And once achieved, the climax ecosystem has the ability to balance fairly exactly its birthrate and its deathrate, its use of materials and its return of them to the soil. It has the ability to absorb stressful fluctuations in heat or cold, rainfall or drought, and can heal rifts and intrusions by reverting to earlier succession stages ("seres") that begin again the slow climb to climax. This is dynamic equilibrium refined to a superlative degree.

Indeed, the actual experience of some forms of climax association can be powerful enough to account for its intellectual and imaginative popularity. A particularly delightful version of the "mixed coniferous forest" in the Sierra Nevada range is the Jeffrey Pine forest, which in its mature form places the human subject in a reality barely touched by the clichéd "cathedral" image. Giant trunks up to seven feet in diameter stand at uniform intervals in deep shade over a quiet carpet of pine-straw and branch-bits. Walking is easy, for the forest is largely clean of undergrowth except for a delicate haze of slender grasses or occasional wildflowers. A subtle perfume, produced by the pine bark, pervades the air. High overhead, the evergreen canopy hides a few birds, whose calls tend to echo and carry. It is a quiet place. Its age is almost palpable. The knowledge that this majestic system is in some way permanent, or at least stable on a scale of hundreds or even thousands of years, makes an impression not soon lost.

The climax ecosystem is a natural system in which little is wasted, in which most life and death occur in a way that helps maintain life, and in which only unusual or catastrophic change can violate the basic stability. Once this climax stability is reached, it will typically change very little, except in the tiny, secret increments of underlying evolution. In spite of the wide variety of accidents and chances that attend the details of a particular forest's beginnings, it will end up in the same climax form, arriving there through a familiar and predictable series of steps. In this sense, the stages of succession and the final climax seem pre-set, or determined, though such goal-seeking behavior is extremely paradoxical (and in fact highly suspect) to traditional methods of science. Clements explained this behavior of the climax ecosystem by hypostasizing it, or presenting it as an abstract entity with a more or less ideal existence. Most contemporary accounts lean towards regarding succession as a stochastic (selectively random) process, perhaps involving a special form of chance effects called Markov chains.

Yet, while the Clementsian explanation of this phenomenon is no longer seriously considered, community succession and climax still are open to a variety of interpretations. Does this strikingly patterned behavior reflect a level of organization at the community level, some form of emergent order that gives the community a real control over its own processes? Putman and Wratten, for instance, consider the possibility to be worth considering. After reviewing the weaknesses in the "constancy of design" (deterministic) explanation, they conclude that "we must for the present reserve our judgment."[20] The climax ecosystem as a real whole continues to spark scientific interest.

As in other cases, the details of the scientific debate do not greatly affect the wider use of such attractive concepts as climax. This image continues to appear in nonscientific writings—as indeed it should, for the scientific question is not whether the condition called climax exists. It clearly does, as a mostly stable characteristic association of plants, animals, and environment. As a myth, this climax formation therefore represents the stability and completeness of nature.

The myth of the climax ecosystem communicates the ecologistic message of unity and connectedness in a different way from the superorganism myth. Rather than collapsing the individual components into a single identity, the climax ecosystem emphasizes a complex unity-in-diversity. It is, therefore, a step more individualized. It speaks to the importance of both diversity and stability, an equilibrium of very real, very separate parts that are functionally integrated.

○ ○ ○

Nature defined as a system in equilibrium presents an alluring model, one which ecologically minded writers and activists have held up as a reference point for social and individual life. The essential gesture of ecologism is an act of deference to the natural world: one adopts an attitude of *seeking* or *listening,* rather than *imposing.* The attentive human creature, with the cacophony of self-generated voices stilled, discerns in the real biological world those guides needed for full and enriching life—an attitude which an early "deep ecologist," Baker Brownell, called "respect for the normativeness of ecosystematic relationships."[21]

Especially potent is this particular principle of ecological balance, derived from the vision of the stable ecosystem. Recycling of cans, bottles, and paper is of course a ready example, however trivial it may seem on the surface. Behind this practice lies the notion that our cultural habits of consumption and waste ought to be replaced with the habit of gaining the most out of raw materials, using and re-using them frugally to ensure enough for tomorrow, and for tomorrow's tomorrow. The underlying notion is that humans must adapt their civilization for the long haul; and this necessitates using only what can be put back into the system for re-use. Ecological thinking stresses that human society ought to imitate the climax ecosystem, and find a way to balance input and output, so that nothing essential ever truly runs out.

The restriction of deadly chemicals in the environment has taken place in large part because of an increased recognition of the importance and ubiquity of natural cycles. Legislation against DDT—the cause of the sixties which Rachel Carson championed—succeeded because of a widespread recognition that this chemical affected the entire natural balance, and that today's gain may be tomorrow's loss.

The balance of nature is *sustainable* diversity; deadly alien substances cannot be allowed to threaten the future.

Equilibrium thinking offers other ways to achieve social ends. To combat insect destruction of crops and plant disease, many ecologists have recommended natural means which use the ecosystem to advantage, instead of warring with it, such as sterile-male release, more effective crop rotation and diversity, and encouraging a pest's natural enemies. These are being tried for both ideological and practical reasons; for the unfortunate fact is that when used in large, continuous doses, chemical pest control loses most of its effectiveness. At the same time the crops are growing dependent on the chemical for protection, the pests are growing immune to it. Chemicals create only short-term gain. Using more doesn't help. "Despite an increase in the use of pesticides from about 100 million lbs in 1947 to about 1.1 billion lbs in 1974, crop losses due to pests [insects, pathogens, and weeds] have not declined but have remained at an estimated 33%."[22] Losses from insect damage alone are even worse: they "have doubled at a time when insecticide usage has increased tenfold."[23] A vivid example is found in the boom-and-bust cotton crop in Peru where, over about twenty years, tremendous initial increases in production were followed by cycles of steep decline, finally levelling off "not too much better" than where they were to begin with—but with the added cost of maintaining the chemical dependency.[24] In sharp contrast is the vision of the stable ecosystem, in which life and death, gain and loss, individual and community, are balanced in a life-giving equilibrium.

Gary Snyder's essays and poems in *Turtle Island* (1974) make eloquent use of the climax/stability myth, expressing the underlying urge for human society to conform to biological norms. A mature society must be like a mature ecosystem. It must value "protection as against production, stability as against growth, quality as against quantity." Some other, epigrammataic pronouncements from the poet's well-known "Four Changes" essay:

> To grossly use more than you need, to destroy, is biologically unsound. Economics must be seen as a small sub-branch of Ecology. On all levels from national to local the need to move toward steady

> state economy—equilibrium, dynamic balance, inner-growth stressed—must be taught. Maturity/diversity/climax/creativity.[25]

Only within a society that has achieved stability can Snyder's values be realized: the equilibrium that nurtures respect for nature, tolerance of diversity, and creative freedom is violently contrasted with the one-way street of pollution, repression of dissent and diversity, relentless growth and competition.

The goal of a stable-state economy is shared by other thoroughgoing ecologistic thinkers. Callenbach's Ecotopians actually achieve it—even if only between the covers of a novel. Their advocates express it this way:

> In the 21st century, the nation that is truly Number One will be the nation that first learns to live on a stable-state basis within the sustainable resources of the planet. . . . We have had enough of expansionist crises and folly. We must seek a new goal; to live modestly and cooperatively and in freedom within the resources of our beautiful planet and within the energy budget set for us by that lovely star, our sun.[26]

This fictional goal is seriously pursued by ecological thinkers, even though it would mean scrapping most of the current world economic system. Possible ways to achieve it are outlined in such books as *Steady State Economics* (1977) and *Toward Global Equilibrium* (1973).[27] The clear values here are those derived from the vision of an unlimitedly self-regenerating system, the climax formation. They emphasize the notion that life must be lived in balance with all other living things: there is no way to steal materials or energy or space from other parts of the ecosystem, or from the future, without committing slow suicide.

Against Progress

The hostility of this vision to traditional Western beliefs, and to prevailing economic practice, is immediately obvious. Perhaps the most vivid element is the opposition to "progress," at least so far as progress is identified as ceaseless economic and social growth. Three kinds of "progress" have led to the present world ecological crisis: unlimited population growth, uncontrolled technology, and ever-increasing material consumption. These violate the ecological values of stability and cyclical regeneration. Therefore they come in for the most scathing attacks.

Naturally enough, the exponential growth of world human population attracts ecological analysis; it is perhaps the best-known ecological fact of life that uncontrolled populations grow until they outrun their food supply—and then they crash. As *The Limits to Growth* observed: "The basic behavior mode of the world system" is exponential growth of population and capital, followed by "collapse."[28] Biologists all over the country were quick to validate this conclusion, and movements like Planned Parenthood and ZPG (Zero Population Growth) gained in influence and following. To ecological thinking, it seems self-evident that population must be controlled by human foresight, or else it will be controlled by less gentle natural means. The population unbalance cannot be long maintained.

Secondly, technological "progress" is also seen as a real danger to ecological stability. The main targets of criticism are usually armaments and nuclear power plants; both are regarded as antilife, the former for obvious reasons, the latter because of the radioactive waste problem. Technology is seen as uncontrolled wherever the momentum of industrial change seems to be running itself *for itself.* The ecologically minded critic will typically insist that other values than mere newness or raw power be considered in making the decision to accept a possible "advance." Pollution and disruption of the biosphere are usually the main considerations; along with these may be concerns about the quality of human life—spiritual, communal, and individual impacts. These too are part of the balance.

Yet the ecological movement has also celebrated technology —*in a different style.* This is "appropriate technology," the use of machines and ingenuity to reach human ends without disrupting or overburdening natural systems. (We will examine appropriate technology in more detail in Chapter VII.) It might be thought of as the natural expression of net-cost or whole-systems analysis. Snyder again: "What we envision is a planet on which the human population lives harmoniously and dynamically by employing various sophisticated and unobtrusive technologies in a world environment which is 'left natural' "— "a scaled-down, balanced technology."[29] A biologically "fit" technology which gets the job done without displacing too much of the surrounding system may often be more limited in scope than the traditional technological answer. A familiar saying suggests, "It doesn't make sense to split atoms just to boil water." Perhaps one need not spend billions of dollars in atomic reactors, long power lines, and corporate bureaucracies to do what a simple rooftop solar collector will do.

However, the solar collector *is* a form of technology, and may incorporate highly advanced features. Or it may not. The criterion that governs all ecological technology is "fitness": only what can be readily absorbed by the biosphere; only what accomplishes its goal without hidden costs elsewhere that are not justified; only what ensures the long-term survival of both human and ecosystem. As in other ecological questions, the issue of technology is reconsidered in the context that matters—the natural balance. The traditional habit of making technology decisions within an exclusively human context (usually focussing on politics and economics) is seen in this context as a sort of pre-Copernican dunderheadedness. It is flat-earth logic. It ignores what is most crucial: the ongoing ability of the natural world to support a diversity of life in a steady state.

Lastly, "progress" in the U.S. often means *economic* progress, and incorporates both population and technology in a vision of never-ending growth. What party or president has ever questioned the assumption that the Gross National Product must expand every year, and the faster the better? Only the radicalizing influence of ecology has created a standpoint from which this article of national faith can be scrutinized.

"Unlimited growth is the philosophy of the cancer cell."[30] This potent epigram utilizes biological imagery to make a point repeated by many an ecological critic of modern life: that within the biological context, humankind and its products have only two choices—to be a healthy and health-giving *part* of the biosphere, or to be a plague upon it. The image appears repeatedly, as in Snyder's condemnation of "the cancer of exploitation-heavy-industry-perpetual growth."[31] Or related expressions like Loren Eiseley's quoted in the previous chapter: "Is man but a planetary disease?" The ecological worldview rejects the assumption that human acts are somehow in a separate category, not to be judged biologically. Humanity is a part of nature, and no unhealth visited upon nature can fail to afflict humankind, eventually. Uncontained growth of any kind is an unnatural horror. For the balance of nature teaches a different idea.

The uncontrolled consumerism which fuels unlimited GNP growth is therefore the product of a vast and dreadful delusion—"a kind of Addict's Dream of affluence, comfort, eternal progress."[32] The consumer society measures value in the number of material things one owns. "More," "bigger," and "newer" are the only adjectives that matter. The ecological point of view rejects this measure as shallow and destructive. In its place must be a variety of physical, emotional, and spiritual values; these concern *quality* rather than *quantity*. They are designed to preserve the rich and ever-changing equilibrium of body, mind, and ecosystem, not to speed growth and change merely for their own sake. Instead of the "more" of unlimited, cancer-like growth, they offer the "enough" of life-sustaining stability.

The Values of Harmony and Balance

The ecological revolution in thinking bases values on the natural equilibrium. Ecological Man and Woman cherish the diversity of the natural world, and enjoy their participation in it. They enjoy its very limitations, as disciplines that enrich their total life. No way of living that tends to destroy the wider environment can be truly satisfying to them. In addition, they value the "total ecology" of the human—their interior ecosystems of intellect, emotions, dreams, as well as their external bodies—and in this way maintain their personal balance and their place in the system. They respect the diversity of the human species, and regard society itself as an ecosystem that benefits from many roles and many niches. They are tolerant of all attitudes but those that endanger the stability and health of living systems.

They "take a natural delight in diversity, as long as it does not include crude intrusive forms, like Nazi culture, that are destructive to others."[33]

If all this sounds utopian, of course it is. Any way of thinking that wishes to challenge the engrained habits of thousands of years of Western culture had better have a potent vision to offer in its place. For ecological thinking does cross the grain of basic Western assumptions and habits of thought. Western thinking is linear, historical, progressive. It pictures humankind on a vast timeline, stretching in one direction only—from Eden, through earthly trials, to the City of God. One need not be religious to share this thinking; it has produced the modern historical consciousness. Belief in progress is a secular faith on both sides of the East-West conflict. Both capitalism and communism rely heavily on it. Especially in the Eastern bloc, "the future" justifies many an abuse or pain. Whether towards the socialist workers' utopia, or the "great big beautiful tomorrow" promised to capitalist consumers,[34] the assumption is that civilization is on an ever-upward path, ceaselessly growing toward a material utopia. But the ecological worldview values stability over growth, and sees the *cycles* of life as being much more important than the forward-pointing arrow of history. This leads to valuing the present moment and its actual life forms, instead of sacrificing them to the idea of the future.

These, then, are the ways the ideal of natural harmony and balance is expressed in the ecological worldview. It prizes stability over growth or the idea of progress. It conceives of an equilibrium of the human and the nonhuman within one embracing idea of nature. It includes the human spirit as a natural factor. It thinks in cyclical rather than linear terms, and consequently values the present reality more than the theoretical future. It thinks holistically, refusing to focus narrowly where breadth is called for.

To "think like a mountain" (in Aldo Leopold's phrase) one must take the long view, and accept the limits of action within the equilibrium.

Conclusion: Dancing with the Earth Mother

These differences in style of thinking can be summed up in one overwhelming distinction: ecologism refutes the human-centeredness of Western tradition. It refuses to imagine that the human species can "dominate" or "conquer" nature. How can one conquer that which is part of oneself, that to which one belongs? Instead, the best one can do is to cherish the intricate and beautiful balance of the world, and strive to come into greater harmony with it.

The sense of belonging to the natural harmony leads to very different conclusions from those that have typified the most fundamental worldviews of the West. Historically the dominant European religious images have been the angry and judging patriarch-god Jehovah and the forgiving Savior (a logical complement to the Angry Judge, not a contradiction). Both religious visions assume humanity's unique status, and in their classic forms offer redemption from the natural self (the "old Adam," the mortal body, the sin-filled earth) into a higher spiritual self in a divine City of God or heavenly hereafter. Always humankind is encouraged to see itself as separate from nature, as a wholly distinct and all-important spiritual being whose story unfolds upon an earthly stage that is at best unimportant, and at worst a snare and a distraction.

In contrast, the ecological imagination is attracted to non-Western expressions of unity and harmony which draw man and woman into the natural circle of life. Especially the Earth Mother represents this insight. A stronger evidence of the radical break with traditional Western thought could hardly be evinced than this: that the age-old fascination with maleness and power be challenged by traditionally "feminine" imagery of nurturing, belonging, and harmony.

Nature's ability to replace, repeat, and reuse all its components and processes over long periods of time—sometimes vastly longer than human-scale time—leads to the humbling awareness of its peculiar kind of everlastingness. This is not the unchangingness of Jehovah, but a more elusive permanence: a kind of cosmic dance, repetitive, intricate, and lovely. In its complexity it is, if possible, even less comprehensible than the austere and monolithic Unmoved Mover of Greco-Christian tradition. But it is more welcoming, too, for humankind is part of the dance.

Hindu, Zen Buddhist, American Indian, and many other cultural and religious traditions have been raided to help the new/old worldview of ecology express its deeper intuitions of balance and permanence within change. A quote from the American Indian named Sun Bear sums up the many interweaving elements of the notion of ecological harmony. I first saw it on a bumper sticker on a Los Angeles freeway, a fact which ironically underscores the very great practical obstacles that separate our society from these ecological ideals. The bumper sticker read: "Walk with balance on our Mother the Earth." Balanced walking is, if you think about it, not very different from dancing. Both are very different from driving on a freeway.

CHAPTER IV
Cooperation

If an ecosystem is whole, unitary, integrated, then it is an obvious corollary that its parts must work together. It would be absurd for one part of the body to "compete" with another. By the same token, no part of an ecosystem can truly thrive at the expense of another. Individuals, of course, are subject to varying fortunes. But in general, at the level of necessary ecosystem functions, all must balance and succeed.

Every plant and animal on earth has a certain well-defined way of getting its food, protecting itself, and providing the next generation. This is its "niche," its role in the ecosystem. To take the largest possible example, no animal could even exist without plants, which turn solar energy into food. And very few plants indeed could exist in their present forms without the animals that pollinate them and prune them and disperse their seeds and control their voracious enemies. Wherever on earth we find more than a bare one or two species—i.e., just about everywhere but the poles—they are coordinated and coadapted in the web of life.

Darwinism overemphasized individual competition, drawing that well-known picture of ruthless nature. But in fact, competition and individualism tell only half the story of nature. Beneath the surface appearance of competition lie deeply rooted interdependencies and long-evolved arrangements of mutual benefit.

The biologist and writer Marston Bates has put it this way:

> We have got into the habit of looking at the organic world as a mass of struggling, competing organisms, each trying to beat the other for its place in the sun. But this competition, this "struggle," is a superficial thing, superimposed on an essential mutual dependence. The basic theme in nature is cooperation rather than competition—a cooperation that has become so all-pervasive, so completely integrated, that it is difficult to untwine and follow out the separate threads.[1]

How we got into this "habit," and how we are getting out of it, are largely the story of ecological ideas supplementing Darwinian ones.

The excitement and power of current ecologistic ideas is, in large measure, a sense of joyous rediscovery of the harmonious community of nature, a sense which had been blacked out by the spread of a popular and exaggerated Darwinism. The other half of the story—the news that cooperation and organic integration are real—is the nub of the ecological revolution.

Darwinian Competition

Darwin's *Origin of Species* presented competition and struggle as virtually the sole measure of natural relations. This grim vision seems to have been a specialty of Victorian England. As one scholar has remarked: "Only there could Darwin have blandly assumed that the basic unit was the individual, the basic instinct self-interest, and the basic activity struggle."[2] In fact, not less than three eminent Victorians came independently to invent evolution and something very like natural selection: Darwin in 1838, Herbert Spencer in 1852, and Alfred Russel Wallace in 1858. All of them arrived at these concepts by reflecting on the same three points: starvation, population pressure, and the writings of Thomas R. Malthus.[3]

Starvation and population pressure were, of course, becoming all too evident in the new industrial slums of England. And Malthus, the famous political economist, had written earlier in the century of the inevitable tendency of populations to reproduce up to the very limit of their food supply, and hence to guarantee poverty and privation for the masses. Darwin, Spencer, and Wallace saw, often in sudden insight, that if this were so, then surely those who survived would be biologically superior to those who perished. They applied this formula of privation, struggle, and selective survival to the natural world and found a ready answer to the question of species change and origination. Nature viewed through the prism of early capitalist struggle and hardship revealed the immensely powerful concept of natural selection and the survival of the fittest.

Darwin's writing reflects his awareness that his new view of nature would meet with energetic resistance. Most particularly offended would be those trained in the traditional view that the God of Genesis had, in an act of "special creation," made a benign natural world. In order to make his point, Darwin (and many evolutionists following him) overemphasized the cruel and combative side of nature. Darwin did not merely compile examples; he cast the whole of nature into the mold of the hungry wolf, pursuing with deadly fangs the succulent and terrified prey.[4] Contrary examples, though sometimes presented, were virtually ignored by the course of the argument, submerged under the great dominating idea of evolution by struggle and bloodshed and starvation.

Twenty-two years later Darwin would acknowledge the excesses of his original presentation, and offer this excuse: "If I have erred . . . or exaggerated, . . . I have at least, as I hope, done good service in aiding to overthrow the dogma of special creations."[5] But such refinements and second thoughts had little effect on Darwinism as a social phenomenon; the *Origin*'s spectacle of nature red in tooth and claw remained canonical. And as many have commented, these beginnings set the tone of Darwinistic science for many years to come.

A word of caution about this chapter. Though, as the following material makes clear, Darwin's focus on competition has had to be broadened, this should not mislead the reader into overlooking the centrality of Darwin's work to the whole of modern biology. Changes in emphasis there have been, and challenges over this or that mechanism of natural selection. But underlying all of ecology, as under all of biology itself, Darwin's fundamental contributions remain unshaken: that species result from adaptive change over evolutionary time, and that rational inquiry can piece together and understand how life itself works without resort to mysticism.

Natural Selection . . . Revised

In Darwin's nature, adaptation always makes one a better competitor, able to overcome or outdo a rival, usually some less fortunate member of one's own species. Individuals are driven inevitably into combat over scarce resources, with the most severe competition taking place between the closest relatives—those who most resemble one another. Thus warfare is general, and fratricide common or even inevitable. Natural selection continually rewards the hardiest individuals, and the most beneficial variations, by culling out all weaker members. Struggle, victory, and death are the engine that drives a species into variation and continual improvement.

But is fierceness in direct competition really so common in nature? An alternative is seen by Darwin but, typically, not noticed. He mentions in passing the possibility of species tenanting "places which are . . . unoccupied," and admits an occasional "fossil species . . . saved from fatal competition by having inhabited a protected station."[6] But his controlling assumption of universal struggle overpowers these possibilities. He leaves them unpursued.

It would be left for twentieth-century biologists to establish the general principle that, in most cases, successful adaptation leads *away* from direct combat. In 1934 G.F. Gause introduced the principle of competitive exclusion, summarized in the phrase "one species one niche." This idea points to the fact that all creatures do better where direct competition for the same resources can be avoided. Competition pushes species into safe habits. To cite a well-known example, fairly similar passerines (song birds), seemingly occupying the same physical habitat, will on closer examination be found to be organized into well-defined and largely exclusive feeding zones—some species in the understory, some in the upper branches, and so on.[7] Rather than competing, they have specialized to avoid competition. In general, the longer an ecosystem exists, the more it will tend to demonstrate this coevolution or mutual adaptation; the most complex and most evolved ecosystems support great numbers of species, each secure in its niche, rarely or never competing. In terms of energy flow, this makes sense; given time, such specialization enables life to tap more and more of the potential sources of energy. Paul Colinvaux summarizes the idea this way:

> A fit animal or plant is one that avoids competition by adopting some private way of life, its niche, which is all its own. Species result not so much from a struggle with others for existence, but from a process of avoiding such struggle.[8]

Ecologists have further established that most locations offer many more niches than are occupied at any one time—an open invitation for noncompetitive adaptation. It takes long periods of geographical and climatological stability for creatures to become specialized into various niches, and seldom, if ever, are all possible niches filled. There is more energy available, and more ways to get it, than organic life can exhaust. How different a picture this presents from that assumed by Malthusian Englishmen, to whom overpopulation, crowding, and food scarcity seemed so quintessentially "natural"!

○ ○ ○

If competition repels, cooperation attracts. Cooperative behavior confers positive survival benefits. This insight also is present in Darwin, but like the implications of the "niche" concept, remained largely ignored until stressed by later revisionists. If two creatures can perform any biological function—feed, defend themselves, find shelter, or reproduce—more advantageously together, then they will have a survival advantage and be selected for, over those who do not do so. Thus the survival of the fittest is not merely a question of the strongest; it is equally a question of the best able to maximize biological resources by cooperation.

Many kinds of animals show amazing coordination and teamwork in patterns of nutrition, reproduction, and protection within their own species. Vivid examples could be found in almost every school, colony, pride, and pack. In fact, the general good of a group of animals is often served by behavior that seems positively selfless.

"Egotistical altruism" is what the entomologist Wheeler called it; for apparently the individual also gains in some way, by improving the chances of leaving its own genes behind in its own offspring or in that of a close relative whose genes it at least partly shares (a process studied today under the name "kinship theory"). The biological imperative of leaving progeny explains why a mother would sacrifice her life for her chick or cub. Natural selection rewards those whose genes survive.

But the circle of cooperation spreads further yet. Observation of the East African bird called the green woodhoopoe, for example, has discovered the most intricate and elaborate arrangements for mutual wellbeing:

> The birds' social unit, the flock, may have as many as 16 members but only one breeding pair. The other sexually mature but non-breeding flock members serve both as "nest helpers," sharing the burden of bringing food to the incubating female and later to the nestlings, and as "guards," defending the nestlings and later the fledglings against predators and participating throughout the year in such flock activity as protecting the home territory.

Such studies are not mere freaks, either; these researchers report that "more than 100 species of bird share with the green woodhoopoe this pattern of cooperative breeding."[9] Cases like this (and there are many) may go beyond kinship theory, for the genetic relatedness of participants is often far too low to exert any significant influence. Instead, these are seen as examples of "stable reciprocity," the natural tendency of cooperative behavior, once initiated, to replace competitive behavior by offering a more dependable reward to all concerned.[10] And this tendency appears not only between members of the same species, but between species as well. Interspecies cooperation is not the exception but the rule. Most of the world's biomass is involved in some form of mutually beneficial behavior.[11]

Biological research of the last four decades has been rich with evidence for the synergistic effects of associated, aggregated, or cooperating individuals. In the words of W.C. Allee, the student of animal behavior and ecology: "the tendency toward a struggle for existence is balanced and opposed by the strong influence of the cooperative urge."[12] "Urge" here simply means drift or tendency—a built-in lay of the land that, given sufficient time, often allows animal behavior to flow in a cooperative direction. The problems of undercrowding, the physical principles of mass ratios (how much "surface" is exposed to hostile forces), the benefits of social organization, all tend to encourage organisms to find mutually supportive styles of interaction—what Allee calls "unconscious proto-cooperation." Cooperative selection has become a well-established complement to Darwin's rule of competitive selection.

Real Competition and Struggle

Darwin's preferred term for what he saw as nature's general warfare was the "struggle for existence." This term included competition between members of the same species and competition between rival species, as well as the outright combat between predator and prey. This latter case would seem to be an obvious instance of true Darwinian struggle. Yet, even here, ecologists point out that the deadly conflict played out by individuals should also be seen as part of a larger, more benign pattern. Is the wolf the "enemy" of the caribou or deer? Only on the individual level. For, like other predators, wolves are also an essential control on the population of their prey. Without control, herds of browsers like deer and caribou quickly get into real trouble. Their numbers rapidly increase. And their food supply rapidly decreases, as a result of overgrazing. Then not just a few, but very many die. And since mammals take longer than plants to replenish their numbers, in many cases the natural forage of the animals' range responds by overgrowing the range. What next? With all this forage, the remaining browsers are soon in palmy days again, multiplying rapidly, until a pattern of boom-and-bust cycles becomes the norm. Predators, however, function like a governor on an engine; by culling the weak and the extra browsers, they keep the functioning parts of the predator-browser-forage system within acceptable limits.[13]

The history of human "control" of predators such as the wolf and coyote has often illustrated the point. A famous instance, first reported by Aldo Leopold, occurred in the Kaibab forest on the North Rim of the Grand Canyon. The Kaibab Deer population there numbered around 4,000 in 1907; but in 1924 the deer had jumped to 100,000, "coincident with an organized government predator removal campaign."[14] The results in the Kaibab National Forest were what one would imagine: overgrazing, starvation, and so on. While there may have been other factors contributing to this appalling increase, clearly a major cause was the destruction of the predator-check on which the deer herd was dependent. Or perhaps the more accurate phrasing would be: on which the ecosystem was dependent. No part of the system could thrive without the "competition" of predator and deer.

In fact, no elaborate and specialized animals—neither wolves nor caribou—would evolve without the agency of this system of strictly limited, mutually beneficial competition between species. Without the complex reciprocal arrangements between all producers and consumers and predators, only the very simplest could exist. The ecological principle of stability dictates the rather commonsense insight that, over time, a system must either stabilize itself by some means which guarantees the survival of its member-parts, or cease to exist. Hence, over the long eons of life's development, those systems whose parts do not coadapt—disappear. What is left is nature as we know it now: prey animals that reproduce just fast enough to replenish their numbers and feed their predators; predators that are just fit enough to stay alive—but not to overcome their prey completely.[15]

Both the evolutionary existence and the ecological health of prey such as the Kaibab Deer are produced by their so-called enemies. The same is true of the predator. If competition it is, it must be seen as a very *cooperative* competition. Perhaps a better term than this oxymoron is the one favored by ecologists: interdependence.

Parasitism—and Other Styles of Interdependence

The data necessary to develop our idea of nature beyond a strictly competitive and individualistic construct was being accumulated throughout the later nineteenth century. This story forms almost a subplot, a countermovement, beneath the clamorous success of Darwinism. An example is the history of the study of parasitism. Over the course of nine or ten decades, biologists revised their somewhat misleading Darwinistic ideas; the primary agent of this change was increased information that did not fit well into existing preconceptions about the universality of competition and mortal struggle.

Parasitism was, as Charles Singer points out, a subject of intense interest during the years immediately following the publication of the *Origin of Species.*[16] This is not hard to fathom: parasitism exemplifies the deadly and rather horrifying relation between organisms in Darwin's picture of nature. (Darwin's first published treatise was a long monograph on a certain parasitic barnacle, Cirripedia, appearing in 1851.) At this time, all "obligate" relations between two organisms—that is, all associations that were biologically necessary to one or both of the individuals involved—were included under the category of parasitism.

What has been called Darwin's "preoccupation" with parasitism encouraged ongoing work with plant parasites.[17] But results were not always predictable. In 1842, to take an important instance, Schleiden had shown that certain plant roots are always infected with certain fungi. By 1885 further study had surprisingly established that, in fact, some of these plants are actually *dependent* on their fungi for successful germination. This was parasitism of a peculiar sort, quite undercutting the expected victor and victim relation! Similarly, the German researcher H.A. de Bary uncovered in 1860 the interesting composition of the common lichen: that now familiar, permanent intertwining of algae and fungi, each absolutely dependent upon the other.[18] A good parasite story was spectacular and awful, quite in line with the Darwinist style. These new cases of beneficial "parasitism" seemed oddly out of sync with the law of the jungle.

By the 1870s, the categorical term "parasitism" was being supplemented with the terms "commensalism," "mutualism," and "symbiosis"—the latter word contributed by de Bary.[19] To call the mutually sustaining relations of symbiosis a species of parasitism was as awkward as Darwin's categorizing mutual dependence under the general term "struggle." By the turn of the century, both the category and the interpretation of certain cases had become open to dispute. Boelshe, for instance, interpreted the lichen arrangement as "close community" and "brotherly cooperation," in contrast with the alternative view that it constituted algae in "bondage" or "slavery" to the fungi.[20] As Edward Step commented in 1913,

> The time is still not far behind us when every case of constant association of one animal with another was set down as parasitism. . . . Fortunately, there are many present-day observers of nature who are inclined to test the deductions of their predecessors by a re-examination of the recorded facts and a search for additional evidence. . . . A large number of cases formerly considered as parasitism are now known to be something very different.[21]

The "something very different" was cooperation in its many forms. This difference would be striking enough to eventually reverse the accepted terminology; today, "parasitism" is sometimes regarded as a subcategory of the general concept "symbiosis" (or else the two are simply treated separately). According to Marston Bates, at mid-twentieth century: "The prime task of the naturalist is probably to keep parasitism in perspective, as merely one example of the multitudinous relations among organisms."[22] Symbiosis is at least as much a preoccupation today as parasitism was in Darwin's time.[23]

Parallel work on other aspects of plant life brought additional data that challenged the picture of nature as an individualistic free-for-all. In the 1880s, for example, Boussingault (culminating years of work by leading figures such as Priestley, Lavoisier, and Liebig) described for the first time the nitrogen cycle. It was an early and most impressive example of the cyclical and self-sustaining nature of life. The intricate pathways taken by the biologically crucial substance nitrogen wound their ways through the air, sea, and earth, and through the lives of many organisms. And each stage of the journey was a necessary one, a linkage that bound living beings to each other and to their environment. That one stage in this journey involved a supposedly "parasitic" root-fungus only made the point more pointed: life worked in mysterious and unexpectedly beneficial ways.

Discoveries like these began to draft an increasingly detailed map of the connectedness of life. Scientists explored more and more of the many branching paths by which one organism both depended upon and benefitted many others. The growing understanding of cooperation and symbiosis in specific cases, along with recognition of a seemingly universal interdependence, created the conditions for a reinterpretation of the orthodox positions of traditional Darwinism.

An 1882 encyclopedia article by the respected biologist Patrick Geddes declared that:

> A restatement of the theory of . . . evolution . . . in terms no longer primarily of strength and competition, of hunger and battle, but of love and cooperation, of sacrifice and pain—would involve, no less fully than was the doctrine of struggle for existence, a deepened reinterpretation of plant and animal life, and would similarly extend into other fields than those of pure biology.[24]

Geddes shows his awareness that the conclusions of biology would not remain limited to the laboratory. They would inevitably be taken into the wider cultural life as part of the ongoing struggle to redefine humanity and its place in the world. Hence it is hardly surprising that this call for "reinterpretation" of competitive Darwinism was answered by quite a number of Anglo-American and continental writers and scientists. A.R. Wallace himself worked on this problem vigorously, moving far beyond standard Darwinian positions (not altogether scientifically) in his 1889 *Darwinism.* The Russian nobleman Peter Kropotkin's series of essays in the 1890s, later collected as *Mutual Aid* (1902), also went far in publicizing this other side to nature. These and many others sought to redress the Darwinian unbalance.[25]

The Symbiosis Myth

Symbiosis exemplifies biological cooperation. Technically this word refers to any strong or habitual arrangement between two individuals. But the most common usage limits it to the nonparasitic forms: "mutualism" (both sides benefitting) and "commensalism" (one benefitting, the other remaining unaffected). And further, the term usually indicates that the beneficial arrangement is virtually obligatory; at least one of the creatures has come to depend upon it, and very often the two have evolved together to become literally inseparable. Like the stories of the superorganism already looked at, the fascinating arrangements of symbiosis are often presented in a representative fashion; they are taken to reveal a fundamental aspect of all nature. In this respect symbiosis is another of the ecological worldview's major myths. It expresses the natural world's cooperative interdependence.

The most famous and often-repeated example is that of de Bary's humble lichen. In North American forests lichen is commonly seen growing on rocks as a green or yellowish or red splotch that looks a bit like spilled paint. As we have seen, this is not one creature but two which have become inextricably linked: a fungus, which provides physical structure and support, and an alga, which is rather shapeless by itself but, provided with this platform, happily photosynthesizes food for them both. This *ménage* has become a favorite biological parable with a well defined moral; as Peter Kropotkin put it, "Don't compete!—competition is always injurious . . . Combine—practice mutual aid!"[26]

In fact, the world of very small plants and animals abounds with symbiotic connections. An arrangement similar to that of the lichen places an alga within each of the tiny polyps which make coral reefs. Still another remarkable creature is even named the *symbiopholus*, for reasons which Lewis Thomas (always on the trail of this topic) makes clear; it is:

> a species of weevil in the mountains of northern New Guinea that lives symbiotically with dozens of plants, growing in the niches and clefts of its carapace, rooted all the way down to its flesh, plus a whole ecosystem of mites, rotifers, nematodes, and bacteria attached to the garden.[27]

There are plentiful and often-told examples from the larger plants and animals as well. The shark has its pilot-fish, as the rhinoceros has its bird. It is probable that in grassland ecosystems such as the African savannah the grasses have evolved along with ever larger grazing animals. The animals have developed into a pruning system to keep any taller plants from overshadowing the grasses, while the grasses, with their unique capability of growing from the bottom up, have provided the ideal continuous crop for the grazers. Even more closely co-adapted are various species of South American *Acacia,* each of which has its very own species of ant to live within its hollow thorns and fend off potential enemies. To attract these tenants, the *Acacias* have evolved "small, easily detachable structures of high nutrient value" on which the ants feed."[28] In fact, symbiotic relations between plants and animals or insects are practically universal. Paul Feeny remarks that, in general, "every insect or plant population may be said to coevolve . . . with everything around it that is capable of evolution."[29]

Birds and insects have proved especially adept at locking onto specific plants for their mutual benefit. "The number of plants (myrmechores) which depend on ants to carry their seed is astonishing," reports Friedrich Schremmer.[30] Other plants (zoochores) may use animals to do this work. For instance the Australian mistletoe bird, which eats mainly mistletoe berries, has the odd habit of wriggling itself against tree branches during its frequent defecation, thus optimally emplacing whatever berries it has not digested.[31] This is the mistletoe's way of dispersing its seeds; that the bird gets a meal out of it does not hurt. Both species gain by the arrangement.

In fact it often develops that in an interaction like this, the plant seed loses the ability to germinate without passing through the gut of its mobile partner. Such "endozoochores" include the silvery berries of the Western Juniper encountered in the Sierra Nevada, and in Africa (by David Attenborough's account) some of the favorite food plants of browsing elephants. For uncounted ages the elephants have stripped these plants clean, seeds and all. Some of them have naturally enough come to armor their seeds within thick protective layers, to withstand the trip through the elephant. "The paradoxical consequence has been that now, unless the rind is weakened by passing through an elephant, the seeds are unable to germinate at all."[32] At this point such a plant could hardly afford to "protect" itself from foraging elephants!

Sometimes even highly specific symbiotic dependencies can include more than just two participants. In the deserts of southwest North America and Mexico, a single species of moth *(Pronuba)* is the sole means of pollinating the agave yucca. In an elaborate process involving special equipment evolved for the purpose, the moth carries pollen from one plant to another and makes sure the stigma is well fertilized. The female then lays her eggs in the yucca's ovary. Here they will eventually hatch, and as tiny caterpillars find an immediate supply of ripening yucca seeds. This is a clear enough, and far from unique, dependency of plant and insect pollinator. But there is a third participant, the so-called "agave" or ladder-backed woodpecker. The woodpecker nests in the tall dead stalks of the yucca and characteristically eats whatever *Pronuba* larvae it can find. In doing so, the bird controls their number and ensures the survival of at least some yucca seeds, thus guaranteeing itself a future supply of stalks, the yucca future offspring, and the yucca moth future nursery sites.

Symbioses like these are common enough to fill books with strange and fascinating examples. The ant and the *Acacia;* the elephant and the thick-rinded seeds; the mistletoe bird and the mistletoe; the yucca, woodpecker, and moth; all are locked symbiotically not in the dance of death once ascribed to nature, but in a dance of life.

The fact is that all parts of an ecosystem must dovetail; this is what makes it a system. Looking closely at almost any part of an ecosystem is bound to produce examples of close connection and coadaptation. Over time, animals and plants coevolve amongst and between each other, shifting their niches and roles to maximize gain and minimize competition. The symbiotic forms are merely the most extreme examples of this universal tendency toward interdependence.

❍ ❍ ❍

The moral implications of this cooperative view of nature form an integral part of the meaning of "symbiosis," as it is used within the ecological worldview. Just as Geddes foresaw, the reinterpretation of nature carries weight far beyond the discipline of biology. The moral tale of symbiosis has become perhaps the most familiar of the ecologistic myths. A tellingly clear form, for instance, appears in the college textbook on ecology by Eugene Odum. Following the section on various kinds of predatory and cooperative interactions, the author remarks:

> The "lichen model" . . . is perhaps a symbolic one for man. Until now man has generally acted as a parasite on his autotrophic environment, taking what he needs with little regard to the welfare of his host. Great cities are planned and grow without regard for the fact that they are parasites on the countryside which must somehow supply food, water, air, and degrade huge quantities of wastes. Obviously it is time for man to evolve to the mutualism stage in his relations with nature.[33]

The contrast between the Darwinistic competitive interpretation of nature and the ecological cooperative interpretation is partly a matter of focus. If one looks primarily at individuals, one sees competition. If one looks at groups and group interactions, one sees integration. In fact, the Darwinian view is *incorporated* in the ecological, not really contradicted by it. The group effects *include* the individual effects. Kropotkin, for example, never denied the operation of competitive, individual struggle. He merely stated that it was not the only factor operating in natural selection. And in the years since he wrote, exactly that point has been proven. The atomic or individualistic level is only one level among many. In the words of W.C. Allee, "At whatever level one begins, and whatever the point of view, one must study all possible unitary levels before coming to a full understanding."[34]

If there is natural selection at higher levels—the group, society, and community—it potentially counteracts the individual struggle for existence. The selective advantage of group cooperation may outweigh individualistic struggle. If in fact natural selection is operating simultaneously on all levels, from the gene to the individual to the local population to the ecosystem, then the struggle of the individual to survive may well be subjugated to contrary demands and selective pressures from a higher level of organization—the need for the group to survive, or for the ecosystem to persist. The net result, as we have seen, would be that the cooperative necessity would submerge the competitive and individualistic one. How else could such behavior as that of the honeybee's suicidal sting be explained, which sacrifices the individual for the greater good?

This kind of so called "altruism" is, as we have seen, very common in nature. There are countless other examples of behavior that either sacrifices individual good for the good of the larger unit, or declines to take private advantage of a mutually beneficial situation. Only by looking at the net effect can the plusses and minusses of selection on many levels be understood. The selfish survival drives of the gene and the individual are restrained, overridden, or transformed by the cooperative mechanisms which create symbioses, families, hives, communities, and ecosystems.

○ ○ ○

In the contemporary tide of "individualistic" methodology in scientific ecology, the concepts of coevolution and selection on levels higher than the individual are hotly disputed, though in neither case wholly discarded. There is a range of debate among the specialists, while many ecologistic writers have a natural orientation toward the holistic approach to any issue.

Regarding coevolution, those predisposed toward the possibility of higher-level systematic interaction are open to it. Putman and Wratten write that "coevolution does not exist just between pairs of species, but between all the members of an ecological community."[35] Begon, on the other side, is doubtful, calling coevolution an "enticing" idea, but one for which "hard evidence . . . is more or less non-existent."[36]

The possibility that Darwinian selection might operate on a higher level of organization than the individual organism is also in doubt—some would even say, is disproven. The accepted orthodox position is that Darwinian selection operates upon the individual organism. Stephen J. Gould remarks that this idea has sustained two major challenges, one each from the reductionist and the holist directions.

The reductionist proposes that individual genes are the real subjects of selection (that is, survival and self-propagation). Though this is for many still quite a live option (publicized in Richard Dawkins' book *The Selfish Gene*)[37], Gould pretty thoroughly dismisses this challenge as unconvincing to most biologists because of its extreme downplaying of the importance of those levels most biologists spend their lives studying—whole organisms and the interactions between them. On the side of "group selection," Gould draws a tempered conclusion:

> The dust has yet to settle on this contentious issue but a consensus (perhaps incorrect) seems to be emerging. Most evolutionists would now admit that group selection can occur in certain special situations. . . . But they regard such situations as uncommon.[38]

Gould's assessment (originally written in 1977) vindicates the orthodox, individual-selection position, but leaves the door ajar to more exotic higher-order effects.

Some weighty philosophers of science insist that higher-level selection and evolution are inconsistent with neither Darwinism nor scientific method. William C. Wimsatt states outright that there is no reason not to accept the possibility of selection and evolution "at various levels of organization," up to and including "even ecological communities."[39] Elliott Sober concurs, holding that, when correctly defined, *both* the individualistic and the group selection positions are reasonable alternatives, neither excluding the other. The interesting question, he says, is rather the relative influence of these "two forms of selection" in specific cases.[40]

A fair conclusion from the perspective of ecologism would be to declare that the cooperative aspects of ecological interactions are reasonably well proven, and that forms of proto-cooperation or mutual accommodation can be found at a variety of levels, from coevolution of paired species up to system-wide coevolution, evolution, and selection.

The Era of Cooperation

As Odum's words about the symbolic weight of symbiotic relations imply, this nonindividualistic way of regarding nature has the most striking effect on the ways we view human society. The moral of such examples as lichen and *symbiopholus* is a familiar one: everyone profits where confrontation is replaced by cooperation. But making this view of nature's interdependency part of a functioning world-view is not a simple matter, however appealing it may appear. For in a capitalist society, too much cooperation is (traditionally) highly suspect.

The American social and economic systems are based on adversarial relations. The checks and balances of our eighteenth-century Constitution are not patterned after the reciprocating roles in an ecosystem. They resemble more the mistrustful truces achieved between armed camps. Similarly, our economic system has traditionally assumed that each worker is the natural enemy not only of owners but also of other workers; all are competing for the same scarce resource (money). It is true that the "invisible hand" which Adam Smith envisioned regulating the market bears a close resemblance to the self-adjustment of natural systems. The market can quite properly be viewed as a cybernetic mechanism. But this view has never seriously threatened the prevailing assumption that the economic sphere, like the Darwinian jungle, must entail gross privation and suffering, and must rule out any cooperation between members. It is well to remember that during the high point of this Social Darwinist view (the later nineteenth and early twentieth centuries), labor unions as well as corporate monopolies were opposed as contrary to the rule of competition. Solidarity and mutual assistance were regarded as sentimentally appealing but unnatural and unworkable.

Yet it is very clear that in the twentieth century our real practice is not (if it ever was) capitalist in this sense at all. No trend in our social structures has been more unmistakable than the gradual shift towards stricter and stricter controls on the freewheeling, devil-take-the-hindmost competition of old-time capitalism. The common name for the present system is "Welfare Capitalism." This is an uneasy mixed-breed descendant of the original. It attempts to retain the incentives which motivate entrepreneurs and workers. But it eliminates the most extreme penalties of economic failure, as well as the most extreme rewards for economic success. In other words, it fences in economic competition within a wider set of concerns.

It is no longer widely regarded as beneficial to let faulty or dangerous products be weeded out simply by consumer trial and perhaps fatal error. The Darwinian competition and survival of products is held to be less important than other values in this case. Likewise, it is no longer regarded as necessary to let the economically unsuccessful simply starve (as the classical views derived from Malthus and Darwin once held). Social utility has been given a different and broader definition; a person is generally regarded as more than a money- or goods-producing unit. In effect the rationale for the entire system has changed from progress and productivity (with the side effect of prosperity for the "fit") to the general well-being of all. Such a change is possible because the underlying assumption—that all individuals are natural competitors—is being replaced by an assumption that, within an economic system, everyone's work ought to benefit everyone else, and that cooperation is in many cases to be preferred to competition.

Naturally such a change is never complete and never tidy. Competition still exists and is still valued. But it is more and more often bracketed within a wider concept of social benefit. All three branches of government have their ways of intervening when competition seems to be getting out of hand, or tending to benefit only a few. And even Reagan-era individualism and business orientation have not seriously diverted this changing tide of belief.

A strong, though subtle, proof of the enduring strength of cooperative and ecological values can be found by looking at the way in which the Reagan administration has tried to undo them. For instance, the James Watt/Donald Hodel Department of the Interior has overseen a huge offensive against the environmental protection laws of the seventies. Yet the rhetoric of this offensive has been deceptively ecological in language. Watt habitually talked about "stewardship" and "protection" as the rationale for his actions. Hodel, as we will see in a later chapter, has presented headline-grabbing proposals to restore the Yosemite valley called Hetch Hetchy, even while more quietly urging the exploitation of many other treasures. In other words, they were obliged to adopt the language of environmentalism, even while they attempted to subvert it. In the same way,

Reagan-administration cuts in benefits to the poor were always disguised under a heavy blanket of public verbiage about the President's firm commitment to an intact safety net. And even the vast military buildup was accomplished under the paradoxical cover of seeking peaceful negotiations—"through strength." Whether the arms-reduction treaty finally achieved in 1988 was a vindication of this policy or an abandonment of it in the face of an upcoming election is a matter for discussion elsewhere.

In other words, the cooperative values control the discourse. They define the questions and the terms of discussion. Even when they are being cynically manipulated and betrayed, they still define the basic sense of right and wrong, the basic set of American goals. Legislative defeats under these conditions are only temporary. The Reagan program won many battles but has never had a chance of winning the war: it has not been able to undo the fundamental American commitments to the environment, to collective attempts at relieving poverty, and to a cooperative solution to the arms race.

Of course, America is far from knowing how to carry out these cooperative values: the poor remain poor under any president, the environment continues to deteriorate, the armaments pile up. Yet the Reagan revolution has not succeeded in changing the at least theoretical commitment to these values. American society will continue to look for ways to realize them.

The conservative social critic and theologian Michael Novak (a 1983 Resident Scholar at the American Enterprise Institute) has examined the falling popularity of traditional capitalism. It is based on "selfishness," he says, and belief in human "corruptibility." Now this may not be a bad system—if its view of human nature is true. To Novak, it probably is true, since it works. Nevertheless, it is a dreary, pessimistic way of seeing the human condition. And it can hardly compete with social visions that speak to the idealistic or hopeful side of human nature. A system like socialism which calls for cooperation and trust "offers a holistic vision of the self in society, gives history a point, and establishes before the human heart the image of a non-alienated and brotherly way of being."[41] In this, the conservative analyst Novak finds himself acknowledging the potency of radical criticisms of capitalism, such as that offered by the group of Maryknoll priests who in 1972 called for "an alternative system based not on competition and the profit motive, but on cooperation and solidarity."[42] These values have become commonplace in a society that increasingly sees itself as basically interdependent.

Conclusion: Cooperation and Ecology

The shift in American and Western culture towards a more closely knit and cooperative society is by no means caused by ecology or the ecological worldview. This is a social direction which would have been taken even if ecology had never made it out of the biology books and into the streets. The increase of population, the intensity of urbanization, and the complexity of technology all demand it. No such densely organized, complicated culture as ours could possibly work on the mythical ethos of the Old West, the who-needs-you brand of individualism.

But the ecological worldview *is* a part of this general movement. It has gained momentum from it, and given it reinforcement in return. The vision of nature as a realm of cooperative interaction and deep underlying interdependence provides a ready model for understanding our developing society. In the same way the Darwinian view of the competitive jungle "made sense" to Victorians, the idea of the ecosystem "makes sense" to a modern Westerner. It shows him or her how many parts work together. It offers a morally and emotionally pleasing explanation of how one's economic and social existence relates to other people's. It demonstrates that cooperation is a deeper reality than competition.

The image of the symbiotic relationship says all this in a nutshell. More than the superorganism or the climax ecosystem, symbiosis retains the vivid distinctness of its members. It communicates an important moral point by acknowledging that, although individuals may be part of greater wholes, they are, simultaneously, separate beings. It is still *possible* to compete—possible but usually destructive.

Like the other myths of ecologism, the symbiotic pair is another variation on a single theme—the fundamental oneness-in-diversity of nature.

CHAPTER V
Cybernetics

Science looks for causes. Scientific logic depends on discovering repeatable sequences of causes and effects. This method, as we have seen, has led to a strictly materialistic view of the world in which all significant events must be physical, since these alone can create scientific causation. Mind and purpose cannot play a role in this scheme, because they are not located in the crucial chain of material cause and effect.

To explain the way organisms and environments are suited to each other, the traditional, prescientific approach from Aristotle until Darwin had been the notion of purpose. Aristotle explained the fitness of organisms with the concept of "entelechy" or "end"; it was in the nature of things to have a correct role or purpose or goal for which they were made. In Christian hands, this concept easily became Divine Purpose. The *telos* or end defined all beings and placed them where they belonged. And by running this logic in reverse, the existence of well-suited organisms and environments could be made to prove the existence of a benevolent creator.

The Darwinian revolution cast out such divine agents from the study of biology, and declared that immediate causes, not final causes, explained the physical world. But the problem of purpose remained. For while the inanimate nature studied by physicists can only be acted on, animate nature, itself, acts; it achieves ends that seem in some sense to be predetermined. Even without assuming a divine *telos,* life seems to act purposefully in many ways.

First of all, organisms grow. An acorn is somehow destined to grow into an oak. Though the individual details are a product of circumstance, its end is a foregone conclusion. There is some kind of purpose or goal seeking built into it. Secondly, organisms maintain their internal states homeostatically, manipulating the stimuli and input they receive to counteract undesired conditions. Thirdly, organisms act purposefully in the common sense. Living things from one-celled specks to highly evolved animals and plants move toward food and away from danger or discomfort. This too is goal seeking: not merely being moved by external forces, but seeking out a desired state. Fourthly, ecology has suggested that many of the characteristically purposeful acts of individuals may also occur in larger units. Ecosystems, for example, appear to grow through ecological succession toward an end state (the climax), and to regulate themselves homeostatically.

Lastly, the process of evolution has, at its core, a sticky problem of "fitness," which looks like purposefulness under a different name. A species adapts and becomes more suited to its environment. Naturally Darwin himself held to a strictly material explanation: adaptation is merely a result, not a purpose or goal. This remains the accepted understanding and is the crux of the Darwinian revolution. Yet many biologists and philosophers since Darwin have continued to struggle with adaptation and fitness, finding there some hidden form of purposeful behavior that is peculiar to life, and not fully explained by strictly mechanical explanations. As the German biologist Helmut Plessner observes, "The emphasis on function, on relation of ends to means, persists."[1]

The usual challenges to the materialist understanding of natural selection come in two forms. The first is the idea, classically formulated by Henri Bergson, that a vital force of some kind drives this process of nature's experimentation, selection, and adaption. This idea accepts Darwin's banishment of final cause from biology: neither God nor nature has any preconceived form in mind for a species. No end is already, ideally, envisioned. The vital power does not draw life, but *pushes* it blindly through time animating its endless "extemporizing," in R.G. Collingwood's phrase.[2] The fundamental idea here is that the inanimate world lacks this ability to act. Action (and not mere reaction) is the root quality of life, and it cannot be ascribed to a machine or a machine-like process: machines cannot grow or develop. Obviously, though preferable to old-fashioned teleology, this explanation is problematical insofar as it reintroduces a mysterious invisible force, the *élan vital,* which cannot be empirically experienced or measured, and which therefore cannot be of use to science.

A different kind of challenge to Darwinian adaptation is simply that it is tautological: that the explanation lacks substance because it merely restates the obvious. What survives is fit; and what is fitness? Survivability. Because the two terms are locked into a logical circle, these critics (often either vitalists or creationists) claim the explanation is meaningless.[3]

Though this critique may raise an interesting problem of logic (to which we will return in a later chapter), its general criticism can hardly be weighty: few explanations in science have ever been so fruitful as natural selection. In the decades since 1859, literally the whole array of living phenomena has been systematically examined, organized, and understood. Right or wrong, such an explanation cannot be trivial.

Thus the appearance of purpose in evolution continues to lure the unwary into various vitalisms and entelechies, in plain contradiction to Darwin's engrained materialism. But in this, the behavior of the living world is at least consistent. The problem of purpose, as we have seen, recurs at all levels from the individual cell to the thinking person. And in all its forms, it is certainly far from fully explained either by old-fashioned vitalists, or by old-fashioned mechanists.

Purpose puzzles the scientist because it upsets one or both primary attributes of a normal cause-effect sequence: that a cause must be material, and that it must precede its effect. If desiring a certain future state causes it to happen, then this cause is invisible and immaterial, and therefore a violation of scientific canons. But if the "cause" of a biological act is instead seen to be the desired future state itself, then it creates the apparent contradiction of putting the effect (the behavior) before its cause (the goal). To traditional science, either solution is nonsense.

Biological life runs afoul of physics-style logic in another, related way. The second law of thermodynamics, or the law of entropy, states that energy tends to degrade into ever more spread-out and unusable forms. An intensely hot point like a lit match will lose heat energy in all directions, until at last it and its surroundings are pretty much the same temperature. In other words, the orderliness and complexity represented by a distinct, tight grouping of excited (hot) molecules (the match head) will inevitably run down toward a state of maximum disorderliness, with no organized and distinctive groupings. Newton's "hard, massy particles" behave this way pretty reliably.

But life does not. Life continually ignores the law of entropy, not only *maintaining* its order and complexity but actually *increasing* it. The evolution of life from simple to complex forms presents a startling exception to the otherwise universal law of entropy. So does the development of stable ecosystems. Both show living things moving against the flow of thermodynamic decline, generating and maintaining vast complex systems. Of course, life only holds up the entropic flow in isolated portions of the overall energy budget. Taken as a whole, and over the long life of the universe, no doubt entropy is as inevitable as anything could be. But wherever life is, it opposes the entropic tendency. If not strictly a contradiction, it is certainly a (temporary) exception—and a highly interesting one, to say the least.

These increases of complexity create new and unprecedented structures of order. Life is continually introducing genuine novelty into the universe. First life itself, a qualitative change from nonlife. Then ever more complicated organisms and systems of organisms. Then (of all things) consciousness. Whereas Descartes had confidently asserted that an effect cannot exceed its cause, living things continually do exactly that, producing surprise upon surprise in a process that gains momentum instead of losing it.[4] A billiard ball universe would seem to be wholly determined from its first instant; everything, as Newton pointed out, would be posited in its original impulse in speed and direction.[5] But life is clearly not predictable in this sense.

○ ○ ○

This little summary shows just how tough are the problems facing biologists. If on the one hand metaphysical causes cannot be used to explain things, and if on the other hand physical causality (as understood by physics) doesn't give sensible explanations for the most important features of life, then this is an impasse indeed. It is, in fact, precisely the dilemma posed by Descartes: the unbridgeable gulf between life and matter, mind and body. The history of science is littered with answers promoting one or the other side of the duality, usually organized under the headings "vitalism" and "mechanism" or, as we have seen, "holism" and "reductionism."

The continuing effort to find a rational form of holism, some means of understanding life as more than the particles into which it breaks down, without falling into the fallacy of replacing empirical science with metaphysical guesswork, has led to the novel approaches and definitions of cybernetic ecology. By using the relatively new concepts of information theory and systems analysis, this branch of science has evolved powerful new ways of understanding the paradoxical qualities of life. Theorists see the ecosystem, the organism, and other living phenomena as essentially information-processing systems whose structure is a specific form of hierarchical relation of parts. This special kind of structure uses energy, processes information, and escapes the immediate tug of entropic decline. It is the mission of cybernetic science to specify and understand how such systems work.

The development of this approach has an interesting history, which can serve as a convenient means of presenting its somewhat complicated main ideas. The most recent decade, however, suggests stagnation and lack of progress in the scientific use of cybernetic approaches to the ecosystem. While some very real contributions remain in place, cybernetic ecology seems to be, at this moment, something of an unfulfilled promise. As we will see, it is not a discredited idea but rather one more or less in cold storage in the scientific community, at least for the moment.

Nevertheless, its ideas are extremely provocative, and they continue to stimulate new ways of looking at life. Much of the ecologistic movement has enthusiastically adopted cybernetic ecology, creating from it a major synthesis of many difficult contradictions of the Western tradition—those dichotomies of life versus matter, mind versus body, biology versus physics, whole versus part, and the like, which we have encountered often in this discussion. Cybernetics has offered a new way around the historic impasse defined by Cartesian dualism. The ecological worldview has taken it.

The Homeostasis Myth

As suggested in the previous chapter, a biological biology would presumably be one with its own logical method and its own canons of proof, and would yield information about such characteristically biological problems as wholeness and purpose. Clements' work took a step in this direction, by accepting large complex units as real entities. Yet the absence of a mathematical explanation of these entities left them seeming somewhat imaginary, even metaphysical. Who could say whether a "complex organism" really existed? And therefore who could say that its apparent purposefulness was real?

In this context, the British ecologist Arthur G. Tansley introduced the term "ecosystem" in the 1930s as another alternative to Clements' climax superorganism.[6] But what was a "system"? At the time, this concept was only just beginning to be rigorously studied. In its own way, it was as problematical a concept as the one it replaced.

The landmark work which began to explore how systems function is Walter Cannon's classic *The Wisdom of the Body* (1932), a book about the human organism's ability to regulate its internal state: to keep its temperature, fluid levels and concentrations, and the like within acceptable limits. Cannon's term for this process was *homeostasis:* "the means employed by the more highly evolved animals for preserving uniform and stable their internal economy." Such homeostatic balance was a uniquely organismic behavior, and it contrasted dramatically with the mechanical balances that could be achieved by inert matter. Biological equilibrium was an active, self-correcting process, unlike "simple physico-chemical states, in closed systems, where known forces are balanced."[7] The point was driven home by striking and detailed expositions of how the body used its various systems—neurological, glandular, cardiovascular, etc.—to counteract or suppress undesired conditions, whether caused by internal or external influences.

This powerful description of organic self-regulation left a highly suggestive word in the scientific vocabulary. Homeostasis was a useful new concept that specified important and little-understood biological behavior. Cannon himself did not hesitate to speculate that homeostasis might well be a concept applicable to other, larger forms of complex organization. In fact, this kind of self-regulation was just what Clements had described in the climax formation. Cannon's detailed work with the individual organism brought Clements' work with the ecosystem a step closer to clear, empirical exposition, for it was apparent that the homeostatic processes in both were similar. And, of course, the individual organism was much less dubious as a real entity.

The concept of homeostasis thus linked several important parts of the biological puzzle together: large units and small ones (ecosystems and individual bodies), and purposeful behavior (self-regulation). It suggested that there was, indeed, something like a typically biological "system" which demanded creative and independent study. Because of this, the concept of homeostasis readily became (and has remained) an important ecologistic myth, presenting an image of biological functioning that could be applied to living things on virtually any level. Cannon himself saw the potentials of this idea for wide application to other aspects of life. He applied homeostasis to other biological systems right up the scale, from single cells to human civilization and social groups.

> Just as in the body physiologic, so in the body politic, the whole and its parts are mutually dependent; the welfare of the large community and the welfare of its individual members are reciprocal.[8]

As Cannon explored the implications of this idea, it is quite remarkable that he came to virtually *all* of what would later be grouped together as the important ecologistic ideas: holism and the superorganism, cooperation, stability. He would not be the last to find the connections of these radical ideas compelling. It was an early glimpse of the deep connectedness of different kinds of biological systems.

Ludwig von Bertalanffy and General System Theory

When Clements and Cannon initially wrote about homeostatic behavior, no one had yet created either the models or the mathematics to allow a scientist to say exactly what was happening when a body or a system "regulated itself." Both Cannon and the Clementsian ecologists could describe systematic action and counteraction as the self-regulating effect took place. But none of these scientists could say exactly how such an effect happened, or why. It seemed to work "automatically"; but as Cannon admitted, "What sets it going we do not know."[9] Showing that it happened was not enough; until the behavior was organized under a verbal and mathematical model, it was not a fully developed science at all, but merely an interesting set of unexplained observations.

What Cannon, Clements, and other investigators of homeostatic balance were forced to do was to regard the regulating system as, in current jargon, a "black box": a "unit whose function may be evaluated without specifying the internal contents."[10] They could study its effects, but not explain its workings.

In this context, the Austrian (later Canadian) Ludwig von Bertalanffy stands out as the proponent of a new way of dealing with such phenomena. In the early 1920s, as Bertalanffy describes it, he was

> puzzled about obvious lacunae in the research and theory of biology. The then prevalent mechanistic approach . . . appeared to neglect or actively deny just what is essential in the phenomena of life.[11]

"Mechanist schemes" of analysis simply failed to answer fundamental questions about organisms, for in biology especially, "we are forced to deal with complexities, with 'wholes' or 'systems.' "[12]

Bertalanffy's classic 1950 paper summarizes decades of his preliminary work in systems. It outlines these major problems of biology, and proposes a Theory of Open Systems to account for them. The theory enables the scientist to deal with those apparently purposeful wholes that elude conventional analysis. And it promises to do so in a quantifiable and rigorously nonmetaphysical fashion.

According to Bertalanffy and the other pioneers of the new approach, the various phenomena of life must not be studied piecemeal, but as examples of a single phenomenon: the behavior of systems. All complex organizations have certain important features in common. An ecosystem, a human being, a mind, a single cell, a social grouping—all follow definable patterns of internal organization and systematic function. All are capable, on different levels, of "purpose," self-regulation, and self-propagation—those unique characteristics of life—because they are integrated whole systems.

> The structural similarity of such models [i.e. these various kinds of system] and their isomorphism in different fields became apparent; and just those problems of order, organization, wholeness, teleology, etc., appeared central which were programmatically excluded in mechanistic science. This, then, was the idea of "general system theory."[13]

Bertalanffy and his school have set out to understand what makes a system a system.

○○○

A work appearing in 1941, written by Andras Angyal, laid down some important basic principles for understanding and defining the system. This was *Foundations for a Science of Personality,* a work of psychology. The ability of a system to maintain itself despite varying material conditions and changes in its physical parts is based on the fact that the system *per se* is not material at all. That is, its identity as a system does not lie in the atoms, molecules, cells, organs, or individuals of which it is made. This recognition was Angyal's major contribution to the problem of finding exactly what a system was. "Wholes cannot be compared to additive aggregations at all," he states. They are literally imcomparable. Systems are "something of an entirely different order." This "different order" is simply the arrangement of parts, their position in relation to the others. "The system is an independent framework in which the parts are placed."[14]

In an organism, the actual materials of its physical body are constantly being changed, replaced, sloughed off. In the same way, a human organization may replace its functionaries, officials, and members. Yet each system remains itself. Each system has, somehow, an essential identity that is unchanged even by wholesale exchanges of parts. This is an old problem in traditional philosophy: what is it that remains constant in a tree or person (or even a stone) amidst the changes and fluctations of existence? Is a certain tree standing in my back yard still the same tree in twenty years, when many of its actual molecules have changed, many branches been broken or pruned, and many others grown? If it is, then what or where is that identity, that sameness? A traditional philosophical answer would be to hypothesize a "substance" more essential than the mere accidents of appearance—a metaphysical essence in which identity lies. Obviously, such an answer is of no interest to science, since it operates beyond the limits of empirical verification.

The answer offered by Angyal is to locate identity in the *relationships* a system maintains among its parts. In the quotation above, Angyal uses the term "framework," but the word does not refer to any physical skeleton. The framework is a "constructive principle" or plan virtually independent of its physical parts. This plan is the "different order" of which he writes, a different level of logical type. To go from the parts to the framework that organizes the parts might be thought of as analogous to the shift from two dimensions to three. The new dimension is indescribable in terms of the simpler; it is literally incomparable. It is not, however, metaphysical in the way a "vital principle" is. The framework or plan exists as an immanent fact of the system; it is not merely "in" the pattern of relations, but rather *is* the pattern of relations. It does not need to exist anywhere else. But the fact that it is a characteristic of the whole, rather than the parts, means that logically it cannot be reduced to any of the parts.

Thus it doesn't matter what electron inhabits a certain energy level in a given atom; it is the *relation* of energy level to nucleus that is definitive. Likewise, as has been often and ruefully pointed out, it seldom matters much which individuals inhabit institutional or political offices; the institution has already defined the office and thereby the office holder. What the office "does" can change little because it is located in a functional web of processes that limits and defines it.

What Bertalanffy calls "the constants of reactions"—the instituted functions and relations—define the system. Its identity is a pattern of relations, an integrated network of processes. Its reality is not of atoms or of basic parts, but of a way of linking these in active relations. In other words, its "substance" is its relations.

To define the real world as process and relation seems a dangerous step away from stable and commonsensical notions. Interestingly enough, it is precisely the same step taken by modern atomic physics, which too has had to let this solid world dissolve, on the atomic level, into a shifting play of temporary forces and relations. This step towards a relational definition of reality also holds great rewards in enabling us to deal with those important larger units, those wholes, which organize and embrace our lives and our world. If systems thinking makes basic matter somewhat less solid—turning it, as modern physics does, into systematic interplays of energy—it also makes the phenomenon of life rather *more* solid. For it clearly establishes the identity and reality of higher levels of organization which physicists have been reluctant to admit into science.

Norbert Wiener and Cybernetics

Angyal's philosophical definitions came at a time when the actual functioning of the system was at last beginning to become understandable. The breakthrough came largely through the work of a brilliant scientist employed during the Second World War by the U.S. Army, Norbert Wiener. Wiener invented the term "cybernetics" to describe a new field of study. The root word is from the Greek for "steersman," and refers to the actual process by which a system steers or regulates itself—"the science of maintaining order in a system."[15] Cybernetics is the science that pries open the "black boxes" to find the controlling, purposeful agent within every functioning, self-maintaining system. Cybernetics, in Wiener's work, would seek to provide a mathematical description of such phenomena as Cannon's homeostasis.

Wiener arrived at his new science by working not with complex, many-layered biological process, however, but with much simpler mechanical and electronic self-regulation. His wartime task was the problem of directing Allied anti-aircraft fire in such a way as to increase the probability of a hit, amidst the randomizing elements of aircraft movement, vibration, wind, etc. To make the problem more difficult yet, the radar images which the guns used were themselves somewhat contaminated by electronic noise and random inaccuracies. Wiener's task, undertaken at MIT (and in collaboration with Bell Laboratories), was to eliminate enough of the "noise" or distortion to get usable information, and to apply that information in a self-regulating gunnery system.

Wiener was an eccentric genius whose writings can be rather technical and abstruse. But the leading ideas of cybernetics can be summarized (at least for present purposes) under two fairly comprehensive headings.

One is that Wiener regarded all events and information as *statistical* questions; not discrete packets of solidly determinate reality, but patterns of data which, when taken as a whole, could have meaning. As Jeremy Campbell remarks in his recent book *Grammatical Man,* a single blip on a radar screen "makes no sense"; it is like a single sound outside of its language context.[16] The blip must be placed in the context of a number of events, which can then be mathematically sifted to eliminate randomness, error, fluctuation. Such a view of information implied a different world than that of Newtonian physics, where each and every atom, particle, or datum was at least theoretically knowable, and rigidly bound to specifically determined behavior. Wiener's view of information was more like the newer theories of the structure of the atom, which pictured an electron not firmly placed in a Newtonian orbit, but wildly fluctuating, and only probably in a given energy level at any one moment.

Wiener's practical application of this statistical or probabilistic theory of information was an important step in the ongoing abandonment of the Newtonian world picture. Statistical information is approximate, not absolute. And a statistical nature would be the same: impossible to know absolutely, and full of strange possibilities and unpredictable quantities. Its actions could hardly be described as universal laws in the Newtonian sense.

The second element of Wiener's wartime task is the one more closely associated with cybernetics: the *feedback mechanism* by which the anti-aircraft guns could utilize that statistically purified information to correct their firing. The following simple diagram, now widely familiar, illustrates the feedback principle.

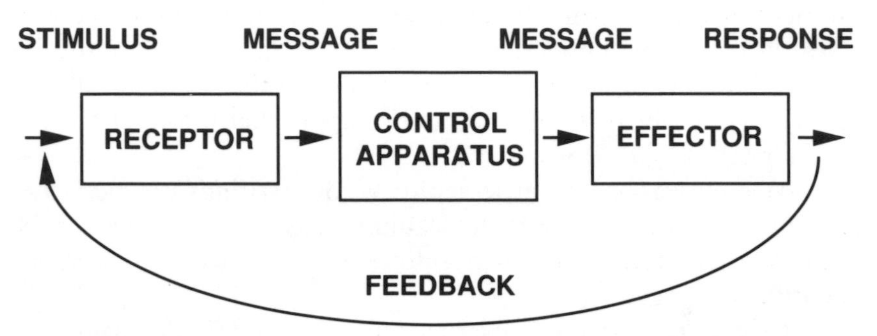

In this schematized version, a sensor or receptor of some type (eyeball, photoelectric cell, etc.) responds to a stimulus by transforming it into a message, which is then sent to a central control apparatus or brain. This acts as a collector which recombines the messages (remember their statistical nature) and sends an appropriate message to an effector. The effector responds to the influence of the initial stimulus. For example, in keeping its balance, a human's inner ear nerve endings send thousands of responses along to the brain. So do the eyes, and whatever body surface is in contact with the ground or other object. All these nerve impulses are processed to identify the body's position; and corrections for undesired deviations can then be sent along to the muscles.

Such a system is called a "negative feedback loop" because the messages record deviations from the set norm of the system and these messages then prompt the system to *negate* the deviations by contrary response. In effect, a negative feedback system "reverses the sign" on the initial stimulus. An overcorrection in one direction will in its turn call forth a recorrection in the other, and so on until relative stability is achieved. This is a continuous closed loop—in Wiener's words, "a method of controlling a system by reinserting into it the results of its past performance."[17]

When Wiener's system was emplaced in 1944, anti-aircraft hits on the German V-1s rose dramatically from a mere 10 percent to 50 percent.[18] His work also provided a clearly conceived, quantifiable model for how living processes might be understood cybernetically. The central idea of negative feedback could be seen in simple, single-feedback mechanical structures like the steam governor, or in almost infinitely ramified versions of multiple, overlapping feedback loops, as in the living organism's maintenance of physical balance or body temperature. The idea was dizzyingly simple, and at the same time remarkably promising in applications to highly complex phenomena. Small wonder that Wiener's *Cybernetics* announced the coming of a "second industrial revolution."[19]

Wiener's cybernetic feedback loop provided an important basis for the then-new science of artificial intelligence and computers. But Wiener also, as Bertalanffy says, "carried the cybernetic, feedback, and information concepts far beyond the fields of technology and generalized it [sic] in the biological and social realms."[20] Wiener had shown, in detail, exactly how information could be used in a system to counteract randomness, error, and entropy. The cybernetics concepts applied equally to machines, to bodies, and to minds, for all could be information-using, self-correcting systems.

This is the basis on which Bertalanffy and others have attempted to build a science, and especially a biology, of systems. Work on systems has given a new identity to the large units which biologists study. The multitudes of cells, organs, and thoughts which make up an organism; the thousands of plants and animals which make up a prairie or a Jeffrey pine forest: they may be regarded, literally and factually, as united on the level of *system*. Within the cybernetic framework, a term like "homeostasis" may be properly applied beyond its original object, the human body, to objects like ecosystems formerly only within the reach of analogy.

Thus the systems approach may be seen as a legitimization of holism. It acknowledges that the integrated whole exists, and has its own unique attributes. And it permits careful, scientific investigation of not only the atomized parts but also the wholes which they make up.

Negentropy

Cybernetics shows that purposeful, homeostatic behavior can be explained by the concepts of negative feedback and information. Systems science generalizes these ideas into a powerful explanatory model for life itself, based on the symmetrical relation between information and entropy.

Part of the riddle of goal-seeking behavior, such as that of a steady-state system, is that it continually impedes the flow of entropy. The system does not allow the general dissolution of order and dispersal of energy which otherwise characterize the universe. Even so simple a system as a candle flame maintains its temperature and energy organization. Naturally, in a larger context it also obeys the thermodynamic law by degrading some portion of the energy of the universe into a lower form. But in itself, and while it endures, the flame also represents a spot of high energy maintenance and order.

Animate systems do much more. The systems of order they maintain are incomparably more complex than the candle flame. And further, they actually reverse entropy (within a limited framework) by adding even higher levels of organization to the universe. Unlike the controlled experiments of physics and chemistry, which may be intentionally closed to the outside environment, so-called "open systems" freely import and export materials and/or energy. An organism or other biological whole is a steady-state open system: it maintains its own internal conditions while in continuous interchange with the environment. This allows it to hold off the effects of dissolution and heat-death, and to impose its pattern of relations on a continual flux of materials. One of the nineteenth century's central paradoxes—life's apparent violation of the laws of energy—is transcended in a single phrase: "Entropy may decrease in open systems." Bertalanffy forthrightly declares that "The second law of thermodynamics applies only to closed systems, it does not define the steady state."[21] Ecologists H.T. Odum and R.C. Pinkerton, after studying the energy use of ecosystems, recommended "a fourth law of thermodynamics applicable to those open steady states which have self-reproduction and maintenance."[22]

An open system, then, overrides the general tendency of energy to dissipate. The open system does so by continuously importing enough energy to maintain system structure and functions. The law of entropy, formerly supposed to be universal, needs to be paired with a complementary law of "negentropy." There is some thing or process which is capable of reversing the slide towards disorder, of negating entropy. That something is information.

It is the introduction of information that reverses randomness.[23] Entropy is defined as randomness—the state in which energy is no longer divided into hotter and cooler aggregates but is simply dispersed. And conversely, information has been defined by theorists since the Second World War precisely in terms of minimized randomness, i.e., order. As such, information is the reverse of entropy.

The notion of defining information by its inverse proportion to randomness resulted from attempts to solve the peculiar problems of messages sent by electronic means. Even a simple telegraph, with its alternating bursts of electrical energy representing dots, dashes, and the blanks between them, could be interfered with by random charges and bursts in the system. When the transmissions became more complex, sending audible words (in electronic analogue) over the wire or through the airwaves, the problem of random interference became more acute. Orderly information patterns are opposed and garbled by randomness and disorder, whether by loss of some part of the signal, or by the introduction of noise.

The question of separating orderly information from background randomness led Claude Shannon (like Wiener a researcher at MIT and then at Bell Laboratories) to formulate some general laws governing the process of doing so. These laws were presented in two famous papers in 1948. First of all, Shannon proved mathematically what is most obvious and yet most puzzling: that, in spite of general disorder and the admitted tendency of things to run down, messages (information) could be made in such a way as to persist, or even (theoretically) to persist perfectly, against noise, accident, and entropy.

In a sense this is hardly news. The very existence of our highly organized universe demonstrates it daily, at least intuitively. The universe is; therefore organization must be possible. By the same token, humans do communicate; therefore it must be possible to create messages and other structures of order that defy disorder pretty well.

But Shannon's formulae explain *how* such a thing is possible—just the question that mechanist, emergent, vitalist, and holist explanations had always found unanswerable. Jeremy Campbell explains the significance of this new theory of information and its surprising correlation with the Victorian idea of entropy:

> Most striking of all, Shannon's expression for the amount of information, the first precise, scientific measure, the first satisfactory definition of this distinctively twentieth-century commodity, was of the same form as the equation devised many years earlier, in the nineteenth century, for . . . the entropy principle. Shannon had set out to solve a specific problem in radio and telephone communication, and the solution he arrived at, by strict, deductive methods, was essentially identical to the formula for entropy that had been established in the physics of Victorian times.[24]

Yet a third transformation of the entropy formula has emphasized again the relation between information and biological orderliness and stability. Mathematical ecologists have given the information/entropy formula one more shake, and created a version of it that is used to measure the diversity and structure of an ecosystem.[25] Though information theory originated in a field far from the science of life, it created a new means of understanding the orderly systems of organisms and ecosystems and how they withstand the onslaughts of time and disorder.

As a self-sustaining system, an organism or other biological whole must use information (negentopy) to counteract decay and disorder (entropy). If it uses information, it must have some symbolic structure for doing so, i.e., a language. And the language of the system, its rules of self-integration, must be structured hierarchically. That is how languages work: they set up overlapping rules of meaning, of varying degrees of generality, that govern how units of expression may be formed and connected. This hierarchic structure gives full individuality to each unit in the whole—each cell, each microorganism, each plant, each animal—while also binding the individuals into a greater totality—the organism, ecosystem, etc. As Bertalanffy observes, "every organic system is essentially a hierarchical order of processes standing in dynamic equilibrium."[26] The next chapter takes a more detailed look at the languages and hierarchies of biological systems. In the meantime, this chapter will conclude by examining briefly what systematic language and hierarchy produce: namely, biological feedback and self-regulation.

❍ ❍ ❍

The most obvious level of biocybernetics is that of the single organism. The homeostatic self-regulation of an organic body is a marvelous and intricate phenomenon. And how much more so, when one reflects on the many different levels of self-regulation existing within each animal or plant; how each cell is also striving to maintain its optimal conditions, as is each organ, each bodily system, and so on. All of these are unified and subordinated to the needs of the organic system as a whole. And, at least in the case of higher mammals, the organic system itself may even be subjected to a higher-order control from the conscious mind. Only the principles of systematic order—hierarchy, syntax, logical type, cybernetic control—can illuminate how all this meshes into a functioning whole.

Larger biological units can also work cybernetically. The reader may recall Alfred Russel Wallace's comparison of natural selection to the action of a steam (flyball) governor. It was a brilliant insight. But with no developed understanding of how systematic self-regulation worked, it was impossible to pursue. Now, however, the application is clear. Natural selection, on the level of the population, is a homeostatic response to environment. It attempts to maintain population density by selecting the correct gene-forms, i.e., those that can survive and multiply. It uses negative feedback to maintain certain optimal conditions on the level of the population.[27]

A species, seen statistically as a gene pool, will tend to clump around the most successful gene formulations—the norm of the system—with various freaks and varieties spread out in diminishing numbers on each side in a bell-shaped curve. If, however, conditions change—say, in a new ice age—then a new norm would arise. A different gene expression would be optimal in the new conditions, and in time the larger part of the population would be grouped around it.[28] The process by which the population adjusts itself is a negative feedback loop; variations too far from the norm receive the ultimate negative feedback—death or failure to reproduce—and the population stabilizes its numbers in the most survivable mode. It is a cybernetically self-regulated system, by which a population "learns" more and more about its environment: its information is tested, sifted, and "hard-wired" into the species' genes by the cybernetic processes of natural selection.

Natural selection, as a cybernetic mechanism of biological systems, does not stop at the level of the single population. Populations do not exist in a vacuum; they live and interact with many other populations, in competition with some and in cooperation with others. The weal or woe of one may affect the fate of many. We have already examined this reality in looking at the well-researched area of predator-prey relations. Here is a clear case of cybernetic regulation on a multispecies level. As was commented earlier, a predator cannot afford to be *too* successful; it will "succeed" to eventual starvation if it too far overmasters its prey. Likewise, the prey population needs its predators. It will feel little or no selection pressure to save every last rabbit or deer from the hungry hunter. The prey population is better off allowing a certain degree of loss. A predator-prey interaction, once begun, has only two possible outcomes: elimination of at least one species, or systematic balancing of both. Just as an unfit individual or gene pattern is removed by natural selection, so an unfit interaction, too, disappears. It is not hard to see that, over time, only those interactions that have the possibility for self-balancing can persist.

An ecosystem results from the natural tendency for such self-regulated interactions to accumulate. "In the evolution and development of ecosystems, negative interactions tend to be minimized in favor of positive symbiosis that enhances the survival of the interacting species."[29] This is evolution at the community or ecosystem level. Just as a single population stabilizes cybernetically around a norm of fitness, so an ecosystem as a whole tends to seek stability and survivability (fitness). In both cases, the general "principle of stability" applies: mutually survivable modes of interaction persist, while destructive ones disappear. Change, of course, is always inevitable. But only positive (fit) change will endure. Hence undisturbed systems will tend to develop, as chance and evolution add new species, toward greater complexity.

One of the disputable aspects of cybernetic ecology has been the relation of this accumulating complexity to the self-regulatory powers of the system. In the 1970s, Eugene Odum would confidently assert that the one enhanced the other:

> Higher diversity, then, means longer food chains and more cases of symbiosis . . . and greater possibilities for negative feedback control, which reduces oscillations and hence increases stability.[30]

As we have seen, however, this may only be true in certain cases, and under certain definitions of the terms. Given the close connection in information theory between energy, entropy, and information, it is interesting that this stability-complexity relation still has some validity when energy-flow through the ecosystem is made the basis of study.[31]

This process of change toward complexity and stability (in the sense that the pace of change slows or ceases) is called ecological succession. While in very simple systems, as for example small aquariums or test-tube infusions of microorganisms, succession may conclude in a few days or hours, in natural conditions development is typically very slow, sometimes taking hundreds of years, for example, for forests. As we have seen, the terminus of succession is the climax phase, the optimum systematic arrangement of all the present organisms. This process of change-toward-stability can be expressed as a general rule of systems: "Any natural enclosed system with energy flowing through it . . . tends to change until a stable adjustment, with self-regulating mechanisms, is developed."[32] "It can be deduced from general cybernetic theory that any system that can adopt different states automatically remains in, or after a time adopts, the most stable of them."[33] One might call this the "natural selection of systems," a cybernetic process that encourages the appearance of ever more complex creatures and interactions.

○ ○ ○

Ecologists have explored various methods for detecting and measuring the stability and information content (negentropy) of ecosystems. Mature ecosystems typically balance their photosynthetic productivity and import of organic materials with their losses by respiration and export. This is the "productivity to respiration" or P/R ratio. While young ecosystems tend to produce far more organic matter than they use, old and stable ones merely produce enough to keep going and replace losses. Another way of saying this is that biomass (the total organic matter, living and dead) keeps increasing until the overall system has created enough structure to support a steady state. The surplusses of young ecological phases continue to accumulate until the ecosystem reaches maturity, at which point productivity is kept to a strict replacement rate. The total biomass of early "weed"-type stages is tiny compared to that of the forest stage, for example. But the early stages are highly productive, turning out proportionately huge quantities of greenery, seed, fruit, and so on; while for its size, the stable forest produces relatively little.

Biomass and P/R ratios illustrate the interesting fact that, through ecological succession, living organisms exert increasing control over their environment. After enough different vegetation stages have occurred, the environment is controlled sufficiently to allow a relatively steady state, in which changes or fluctuations can be minimized. In the words of Eugene Odum, succession "culminates in a stabilized ecosystem in which maximum biomass (or high information content) and symbiotic function between organisms are maintained."[34] Criticisms of this idea have followed the pattern of other phases of the complexity/stability question. In some cases, the ultimate stages of succession may present such a tight interlocking of biota, using or relying on all the available biomass, that they leave little margin for flexibility in response to unusual change. Though the general pattern of succession is undeniably toward greater complexity, stored biological resources, and a cessation of succession, to what extent this guarantees stability continues to be debated.[35]

One of the most significant means of stability created by ecosystems, particularly in temperate latitudes, is found in the soil. This was emphasized in a famous paper by the well-known ecologist J.E. Weaver, student and long-time colleague of Frederic Clements. He described the complex history of interactions that made the American prairie such a widespread and durable ecosystem. "The climax vegetation is the outcome of thousands of years of sorting of species and adaptations to soil and climate."

> Upon the fall of leaves and stems the organic matter of the plant, which has resulted from synthetic activity, is incorporated into the soil. These residues of grassland vegetation have returned more to the soil than the green plants have absorbed from it. Throughout their lives plants have synthesized many organic substances—sugars, starches, celluloses, fats, and proteins. Most of these materials return to the soil when the plant dies. This added organic matter produced by vegetation introduces a fundamental change. The substrate is no longer the former one of mineral matter alone, but now contains stored energy in the form of organic material, and a microflora of bacteria, fungi, and other organisms.[36]

An ecosystem may thus achieve equilibrium by creating an environment for itself and then stabilizing its numbers and biomass to maintain a steady state. If its integrity is interrupted, by bulldozer or plow, fire or flooding, the ecosystem will simply revert to an earlier phase of succession, and reduplicate the slow process of soil and biomass accumulation, until the wound is at length healed. Such self-maintenance and self-regulation, however slow in operation, are seen as homeostatic behavior. To paraphrase the Spanish ecologist Ramon Margalef, the process of succession is equivalent to a process of accumulating information.[37]

Margalef, in fact, goes a step further. He defines the ecosystem as "a channel which projects information into the future."[38] This means that a successful system (of any kind) must be able to limit the number of possible future states it will inhabit. Otherwise, the system will not persist; entropy will pry apart its relationships and mere randomness will replace systematic integration. The more stable a system is, the better able it is to negate imposed changes from outside; its composition and functioning are self-determined. "Any cybernetic system, through the interactions of its parts, restricts the immensely large numbers of a priori possible states and, in consequence, carries information." "In this context, information is anything that can influence and shape the future."[39]

Looked at in this way, the many kinds of ecosystem self-regulation we have already examined can be understood as information circuits that control the future by resisting uncontrolled change. Perhaps the most vivid illustration occurs where ecosystems face regular cyclical threats to their stability. In climates with well-defined seasons, most ecosystems have equally well-defined structures *already in place* that anticipate and neutralize the stress of changed conditions.

This is so familiar that its significance may be lost: this kind of behavior clearly requires some advance information about what is going to happen each winter, spring, summer, and fall. That information has been structured into the ecosystem on every level, from the genetic to the systematic, during the processes of succession and evolution. In the words of Margalef again: "One can say that the ecosystem has 'learned' the changes in the environment, so that before change takes place, the ecosystem is prepared for it."[40]

Information-using systems are defined by their ability to organize and persist, i.e., to be counterentropic. It is information that makes an array of organisms an ecosystem; and it is the use of information that enables an ecosystem to thrive in a changeable world.

The Unfulfilled Promise

The systems- and information-theory approach to ecology has been both powerful and popular, apparently reaching something of a peak in the mid- to late 1970s under the leadership of such eminent scientists as Eugene and Howard Odum and Ramon Margalef. From 1964 to 1974, a cooperative worldwide scientific effort known as the

International Biological Program (IBP) expended up to forty million dollars each year, and drew on the work of thousands of scientists. This program was heavily influenced by systems theory, and much of its effort went to gather "quantitative descriptions of the pattern of flow of energy and matter through the communities"—information which the systems analysts craved. This huge undertaking has created a large factual base for ecologists (of any persuasion); some of the findings are still being assembled and published.[41] But in 1980, a very critical Daniel Simberloff characterized the cybernetic school as past its "vogue," though still "a powerful force in ecology today."[42]

At present, the brave pronouncements of Wiener and Bertalanffy and the bold hypotheses of Odum and Margalef, seem to be the well-wrought doorways to a still unconstructed, or at least largely unfinished, edifice. The approach is not discredited, but is at something of a standstill. McIntosh stated at the end of the seventies that "the promise of Information Theory for biology generally has not yet been fulfilled."[43] Though the productions of a major figure like Margalef are still warmly received, they do not seem to stand at the center of a developing paradigm, as they did a decade ago.[44] Recent textbooks make little mention of cybernetics or information theory *per se,* although many of their basic concepts do appear, typically in scattered references to negative feedback, self-regulation of populations, Shannon's formulae, and the like.[45] The real contributions of this approach endure; but cybernetic ecology is not currently being used as a major organizing idea.

It may be that one reason for this loss of momentum in the cybernetic/information theory paradigm in the biological sciences has been the sheer difficulty of identifying the elements in a dispersed, naturally ocurring system such as an ecological community. The following chapter will explore these difficulties a bit more; the obstacles, however, are considerable. No doubt another liability has been the fact that systems ecology is the heir of the long-embattled holist tradition in ecology. Simberloff, for instance, remarks that Clements' superorganism has been "transmogrified" into the cybernetic ecosystem; I believe this to be true but not damning, since the transformation has been in the direction of making the "real whole" quantifiable and subject to rigorous study. Nevertheless, if such an attempt succeeds, it will be against the hostile opinions of the majority tradition in science. As Robert May comments, systems theory "still awaits its Kepler, much less its Newton"[46]—though no doubt Wiener would disagree. It is still a young science. Time will tell.

Conclusion: The Global Mind

Though further advances in systems ecology appear to be on hold, the broader implications of this developing science have made quite an impact already. The information-based understanding of the ecosystem is part of a revolution that radically restructures our thinking. The universe as envisioned by systems science is a far different place than that envisioned by Newtonian physics. It is a place full of order and creation, full of the generative processes that bring life, innovation, and thought out of the simplest cosmic starting points. The slowly accumulating patterns of relation that build up in star systems and galaxies also build up in organic molecules, cells, organisms, societies, and ecosystems. The universe is made in a way that at least allows—possibly *encourages*—information to clump and accumulate and organize. It has a law of negentropy that matches its law of entropy. This is the surprising news of cybernetics, and the beginnings of an answer to many of the long-standing riddles of biology.

This brings us to the second major myth associated with the cybernetic description of the ecosystem: the notion that "mind" is an attribute not just of human beings, nor even just of animals, but of virtually the entire realm of nature. Steady states can be observed in many places, from the microscopic atom to ecosystems to the whole biosphere. Units both larger and smaller than human-sized organisms may be said to be open, steady-state systems, and therefore to process information, *and therefore in some sense to think*. It is a striking, even a shocking idea. Natural mentalities, information-users, are in fact around us everywhere.

Descartes divorced the category of mind from all else in the universe. Cybernetic ecology reunites them. Humanity's conscious mentality is only one end of a long natural spectrum of self-referential, purposeful information processes. Suddenly, the human mind belongs here, on planet earth, along with the many other strange and lovely intricacies of life. The world within which men and women wake and dream is not alien to them, but is as familiar as their hands, their bodies, their very thoughts. In fact the two are interpenetrating, humankind and nature expressions one of the other.

Where we are then, is someplace vast, language-studded, imagining, perfectly balanced, and beautiful. "The only place this can be is the *Mind.* Ah, what a poem."[47]

The mind of nature is the world's total information processes, from the smallest cell to the largest comprehensive global network called "Gaia." This is the ultimate myth of ecology, the superorganism reinterpreted through information theory and raised to a level that incorporates what is most human—thought—with all the rest of nature. The details and implications of this myth are extensive and perhaps a little dizzying. They are further explored in the following chapter.

CHAPTER VI

The Natural Mind

Cybernetics provides the basis for the ultimate extension of the ecological worldview—an extension that reveals the radical unity of the human species and nature. For recognition that both the human mind and body belong to the endless continuum of nature leads to a comprehensive vision of the cosmos that is perhaps best described as religious. It is not merely that "mind" becomes seen as fully natural. It is equally that the "natural" becomes seen as, in a profound and unexpected sense, fully mental.

The assertion that information-use, language, and mind are bound together in all natural systems is at first a bit hard to swallow. Yet one of this century's most dramatic biological breakthroughs illustrates the connection: the discovery of DNA by James Watson and Francis Crick in 1953, and the subsequent cracking of the code of life. The demon at work in the microscopic core of organisms, busily sorting and arranging molecules and defeating the effects of entropy, has indeed turned out to be information.

We have all heard the familiar DNA story. The elaborate instructions for constructing and maintaining an organism are stored upon the famous double helix molecule, and written out in a simple alphabet of merely four "letters"—the nucleotides adenine, guanine, cytosine, and thymine (called A, G, C, and T for short). At this level there is no question of any simple cause-and-effect or chemical explanation for life processes, as prophesied by mechanists like Jacques Loeb for example, who confidently asserted in 1912 that all life would soon be "unequivocally explained in physico-chemical terms."[1] For the DNA molecule clearly functions *symbolically;* its relation to the fibers, organs, and tissues of the body is not direct, like a photograph or even a blueprint, but indirect. It literally is encoded information, like words on a page.

The comparison of DNA to language is not a mere analogy. For the amount of information stored on the forty-six human chromosomes must be vast. How could it be stored with only four simple chemical substances? And, equally intriguing, how is it possible for these codes to generate not only the familiar—reproductions of forebears, for example—but also the novel or unprecedented variation? For evolution surely demonstrates the importance of more or less continuous "experimentation" by a gene pool, turning out individuals with minor or major differences in just sufficient numbers to respond to a potentially changing environment without wasting too much of the population. How can the four chemical bases A, G, C, and T do all this?

The answer lies in the nature of language—language broadly conceived as any rule-governed system of symbols.

Language and Innovation

Language is literally inexhaustible. Whether in the language of words, the language of clothes, or any other complex system invested with symbolic meaning, the process of use is almost indistinguishable from the process of creative transformation. While in a spoken language, for example, the number of sounds and even words (arguably) may be fixed, the number of possible expressions is infinite, because the creation of meaning is governed by layers of rules and rules-about-rules that allow for the continual production of new expression. No speaker of language merely parrots previously heard sentences; he or she creates sentences.

Because of its dependence not upon set and immobile expressions but upon rules of making, language differs from, for example, the simple communications of animals. Certain colors, gestures, or sounds in the animal world may often have nonliteral significance.

But this significance is fixed, like the meaning of a red stop light; it cannot be manipulated. Birdsong, bright colors in insects, and the like appear to be simple one-to-one signs that announce territory, danger, inedibility, and so on.

True language, on the other hand, is not a set of meanings but a method of making meanings. It relies on orderly overlays of context, and on allowable ways to vary the interaction of elements within and between the contexts. These overlays of context are a form of *hierarchy:* each level operates on its own, but is also subsumed by whatever logical level is above it or embracing it.

For instance, the rules of tone declare that a statement may mean its opposite. One may say, "Well, I guess I don't have enough to do," and by properly ironic tone invert the normal syntactical meaning. The rules of tone radically change how the rules of sentence meaning are applied. In this case the "tone" framework brackets the "syntax" framework, bringing an altered meaning out of the sentence; instead of meaning what it literally says, it means its opposite. These frameworks are themselves recognized by a hearer only because they are placed within wider contexts yet—a sense of the speaker's personality, the immediate social situation, and so on. The creation of meaning in each of these contexts operates not by mere replication of set patterns, but according to rule-governed methods of production. Where every expression, and every rule of expression, may be altered in an orderly way to produce understandable innovation, the possibilities are literally endless. A small rule change creates huge new possibilities. The hierarchical interaction of contexts and rules governing them provides a matrix for the continual production of new expression.

○ ○ ○

As applied to DNA, this understanding of language is particularly revealing. For it is increasingly evident to investigators that the organization of DNA is not simple instructions for making the body's molecules. The body is far too complex for that to suffice. Long sections of the DNA helix seem not to correspond directly to *any* specific molecule-building task. These sections may be simple redundancy, built in as a safeguard against error. But it is also likely that some of these sections may contain the rules by which the DNA is to be read and used—its own grammar. Therefore the probability of further layers of grammatical complexity is open: instructions about when to use a certain type of rule for assembling this or that molecule, as well as instructions about when to use these instructions.[2] In other words, the DNA functions as a true language system, self-referential and as full of untapped potential as any other language.

The four letters of the DNA alphabet would not then be a simple set of signs, like birdcalls, with only one possible meaning in each combination. They would be instead a real language, in which each statement may be bracketed within any of various context-frames and operated upon by any of various rules which lead to appropriate and highly varied output. It is this grammatical aspect of DNA which would enable it not only to construct the vastly intricate machinery of the body, but also to introduce new expressions which are not merely random mistakes but are "meaningful"—i.e., rule-bound, grammatical—in the same way a human speaker may produce sentences that are not mere replications of previously heard sentences. Even if a speaker is making up a new word, he or she will construct it to be phonetically appropriate, and will fit the new word into a familiar pattern of grammatical usage (as a modifier, say, or a verb), so that it will be readily meaningful to the hearer. Only if the speaker is "crazy"—unable to recognize context-frames or apply appropriate creation-rules—will these verbal inventions be meaningless.

Rule-governed innovation, unlike chance mutation, may often produce usable results. In British biologist B.G. Goodwin's words:

> We now recognize. . . . the symbolic nature of the genetic code and the remarkably elaborate system cells have for its translation. It is genetic symbolism that enables living matter to step outside the constraints imposed by physical laws . . . The symbolic nature of the genetic material is what provides a virtually inexhaustible reservoir of potential genetic states for evolution, since symbols can be juxtaposed in very many different ways, to provide new "statements," new hypotheses, which can then be tested.[3]

Or in Jeremy Campbell's words once again: "A modest number of rules applied again and again to a limited collection of objects leads to variety, novelty, and surprise."[4] The ordinary cause-and-effect of the Newtonian universe could only produce expected and predetermined results. But the grammatical universe of information-based, language-using systems is endlessly generative.

In sum, an information-using system is organized on exactly the same principles as a language. Hierarchy, interlocking levels, syntaxes, rules, and metarules—all create a complex organic structure which can refer to and control its own processes. Such systems are their own contexts; they provide the network within which individual atoms, bits, cells, or members take on meaning, function, and life.

It then becomes germane to observe that a system which uses language to make meaning and to refer to itself is doing something we usually associate with the human mind. Is it an exaggeration to see such complex, many-layered functions as mentalities?

Mind and Body

Nature appears to abound with information-using mentalities—minds which are located many places beside the human skull. Central to this way of looking at nature is the strategy of creating a wholly new category that includes both sides of the mind-body duality: the idea of *relation* or *pattern*. Jerry Fodor has surveyed the new approaches, and he comments:

> In the past 15 [sic] years a philosophy of mind called functionalism that is neither dualist [i.e. vitalist] nor materialist has emerged from philosophical reflection on developments in artificial intelligence, computational theory, linguistics, cybernetics, and psychology. . . . In the functionalist view the psychology of a system depends not on the stuff it is made of (living cells, mental or spiritual energy) but on how the stuff is put together.[5]

The phrase "psychology of a system" suggests just where the connection lies between ecology and these developing theories of mind. Within the ecological worldview, systematic interconnectedness is an essential feature of nature. And a "systems view" of mind shows that even mental activity can be included within the scope of purely natural phenomena.

Recent experience with the ever-multiplying capacities of the computer adds a lot of weight to the functionalist position. The idea of imitating the human brain in computer form is an open question to many—some would say it is a genuine theoretical possibility.[6] Naturally, the practical difficulties are immense, far beyond the current state of both hardware and software arts. Nevertheless, if the human neural network can theoretically be replicated in metal and silicon, then one is bound to look for the distinguishing feature of the network called "mind" in some other aspect than the humanity of its bearer, or its biological constitution. Neither "soul" nor flesh appear to be necessary ingredients. What *is* necessary, in the systems/functionalist view, is a number of elements linked in certain kinds of interactive patterns. "The situation seems to call for a relational account of mental properties that abstracts them from the physical structure of their bearers."[7]

To say that mind might be created in a machine is emphatically not to reduce the human to the status of a machine. It is rather to recognize that mind is neither a mystical force, nor a property of matter *per se,* but a result arising from complex patterning. This being so, mind might be found many places outside of the animal skull. Recognizing it would require knowing the typical *function* of a mind, since its parts might be scattered or constituted in unexpected ways. But with the right criteria, one might be able to recognize when an open system is truly functioning mentally. How would we recognize a nonanimal or nonindividual mind? It is not a problem that has occurred to anyone to solve until recently.

Gregory Bateson

A set of criteria for recognizing mind was developed as something of an apex to the life work of Gregory Bateson. During the last decade or so of this remarkable man's life, he found himself elevated to the status of ecological guru (as shown, for example, in the remark of

Stewart Brand—editor of the *Whole Earth Catalog* and *CoEvolution Quarterly*—that "I owe more understanding than I know to Gregory Bateson"). Bateson's last book titles illustrate his progress toward ecologistic affirmation; his *Steps to an Ecology of Mind* (1972) led him finally to *Mind and Nature: A Necessary Unity* (1979).[8] In these works are found some of the clearest statements of the philosophical and religious implications of ecological thinking.

Son of the eminent biologist William Bateson, husband of Margaret Mead, author of a widely read anthropological study of Balinese life, discoverer of a powerful new approach to schizophrenia—Bateson ranged broadly and brilliantly over the globe and through many problems of contemporary civilization. As he spent his last years among Pacific coast redwoods as a teacher and regent of the University of California, his attention focussed increasingly on the holistic thinking of ecologism.

Gregory Bateson's work as anthropologist illustrates the pattern of his thinking: he was typically a synthesizer, a finder of sense and meaning amidst the seeming clutter of events and cultures. Context was his watchword; only by seeing a thing in all of its relations could one get a grasp on it. In his work on schizophrenia, for example, Bateson escaped the artificial limit of the patient's solitary mind, and investigated the *family interactions* that produced the diseased condition. Bateson discovered that a child bombarded by highly pressured, intensely self-contradictory messages could be driven finally to schizophrenia as a means of accommodating the pressures and paradoxes. Some schizophrenia could be seen as a product of a faulty language context, in which the hierarchy of a family's messages and metamessages was jumbled and broken, leading to severe inner context confusion for the child.[9]

As his career matured, Bateson's studies and background suited him perhaps uniquely to attack the larger problems of mind, information, and biology. Drawing on contemporary work in many fields, he concluded that large, patterned systems which are self-regulating and sufficiently complex are in fact minds, systems in which interlocked contexts produce startling results. They are characterized by *hierarchical ordering* and by *self-referentiality.*

The first concept, hierarchy, we have already dealt with briefly; self-referentiality is closely related to it. Every hierarchy presupposes its own lower levels, but in order to be self-repairing or self-replicating it must also include within itself information about how to make and use its own lower levels. Hence, the existence of rules about rules—the metalevels—some of which, as in the DNA molecule, may not be about the outside world or the interface with it at all, but rather about *itself:* how to make, maintain, and use its own structures and inner relations.

Hierarchic ordering implies a kind of reality which does not exist from the point of view of atomistic analysis—it implies the integrated whole. In the same way, self-referential systems imply a type of logic quite different from that applied to the particle/motion problems of physics. The logic of physics is unable to explain context-rich information systems that refer to and organize themselves, for quite different rules are at work there. In Bateson's words, "logic is precisely unable to deal with recursive circuits without generating paradox."[10]

We have already examined one not very complicated form of recursive (self-referential) circuit. Negative feedback is a logic loop that returns the input signal in a reversed form to some mechanism for effecting change. It is furthermore a continuous loop, because the change so created is immediately entered as the next message through the system, and so on continuously until the system is stabilized and the message value very small. Bateson recounts the interesting tale of how an early form of such circuitry—that famous flyball governor used on steam engines—could hardly be made, at first, to exert the proper control. Either the engine would race, or slow to stopping, or oscillate wildly. The engineers were unable to figure it out. What is most notable here is that ordinary analytic method broke down when confronted with even so simple a form of holistic interaction as a flyball governor. It took one of the greatest scientific minds of the nineteenth century, James Clerk Maxwell, to discover the beginnings of the mathematics to deal with this situation.

Positive feedback is another form of recursive loop. One well-known example occurs when sound from an electronic speaker recourses through the system via microphone and amplifier, in a rapid crescendo. The system through which the signal passes contains no means of controlling the augmentation which occurs in the amplifier, and the result is out-of-control escalation. The international arms race can further illustrate the vicious cycle of escalation, response, and increased escalation that typifies positive feedback. The repeatedly re-entered message simply accumulates force, since its sign (plus or minus) is never reversed.

Such loops involve very simple quantitative transforms of information—all they do is increase the force or change the sign of the original message. But when logic loops are integrated in sufficient complexity, they are capable of returning much more elaborate and surprising results. Results can be compounded by hierarchic ordering, yielding messages about messages about messages . . . finally (to take an admittedly huge leap) creating the results we see in living, thinking forms, whose capabilities have for so long seemed inexplicable save by invocation of metaphysical powers and spirits. Beyond the simple recursiveness of self-regulation, these more complicated forms can literally refer to themselves for the various purposes of linguistic context, replication, self-maintenance, and (ultimately) self-awareness.

Self-referential systems are by definition circular: built of logic loops that spiral and turn back on themselves within the dizzying maze of a language system. (Small wonder that Douglas Hofstadter likes to call them "strange loops."[11]) This attribute may provide an answer to a criticism of Darwinian selection which we encountered at the beginning of the last chapter: namely, that it was a tautological nonexplanation. As we have seen, a species can be described as a population stabilized around a norm through negative feedback. Adaptation by natural selection simply reveals this system making use of its homeostatic, self-stabilizing ability to adjust and survive. It is evident that the huge amount of information embedded in a population is, indeed, organized in recursive circuits—just the situation which, Bateson warned, is nonsensical to ordinary logic. The tautological circularity of the critics is in reality the self-referential circularity of an information-rich, complexly organized, self-referential system. It is the circularity that makes living systems so strange, so paradoxical, and so powerful.

The Batesonian Mind

The operation of the complex system of contexts in DNA is not in principle dissimilar to higher order systems of hierarchy and self-referential loops. Weave enough loops together, on enough interlocking levels, and the result is what Bateson terms a "mind."

Bateson's criteria for mind are these:

1. A mind is a whole of interacting parts. This is familiar already as a holistic system.
2. An interaction in a mind system is triggered by a stimulus (a "difference").
3. The energy of response comes from the system itself, not the stimulus. This is a critical difference from Newtonian, "billiard-ball" causation: the system itself provides both the energy and the direction for response. Minds do not merely react to stimuli—they *respond* to them through their own inner interpretive systems.
4. "The mental process produces circular (or more complex) chains of determination." Negative and positive loops are inferred here, as well as more elaborate sorts (self-referential and hierarchically "stacked").
5. Mental systems operate with coded versions of their stimuli. This criterion emphasizes that mental systems transform physical events into *information*.
6. Mind places stimuli and responses to them in hierarchies of contexts, contexts of contexts, etc. All messages within the system are understandable because they are *classified* by higher order metamessages.[12]

None of these ideas is unfamiliar; this list amounts to a summary of basic information and cybernetics concepts. What Bateson does, however, is to combine them under the unexpected title of "mind." Doing so aggressively challenges the traditional divisions between organisms and other systems, between individuals and aggregates, and between living and nonliving. Clear distinctions in each of these dichotomies are erased with the recognition of "mental" activity in special cases within all of them.

In one sense Bateson's definition is based on a relatively straightforward notion: the observation that "mind" is the normal category for systems whose main currency is ideas, or information. He simply does not let the ordinary bias of restricting the term "mind" to humans stop him from applying it wherever it is functionally appropriate. Once defined by Bateson's criteria, however, mentality appears in a surprisingly varied and wide range of places on planet earth. The previous chapter provided some examples, including ecological interactions and the whole process of evolution, in which information—coded upon the genes and stored in populations—is made, tested, and either retained or discarded.

For Bateson, self-regulating information systems turn out to be the common denominator of life itself, and many of the peculiar phenomena of biology can be included under the umbrella of "mind": "thought, evolution, ecology, life, learning and the like occur only in systems that satisfy these criteria."[13] Conscious thought, it should be noted, is only one of the mental phenomena here, for in this usage "mind" does not imply self-awareness of that sort (though self-referentiality *is* a necessary attribute of mental systems, and may be seen as a primitive form of consciousness).

The stable ecosystem is also a Batesonian "mind." We have already examined how such a system regulates and maintains itself. We have seen further how it may be subject to selection pressures on many levels, from the individual to the entire system; and how these pressures tend to create many-layered hierarchical structures of feedback. The group mind of the ecosystem is what enables it to develop into greater complexity and to persist.

The massive mind of the ecosystem may even be seen as part of a yet larger mental system embracing the entire biosphere, a global system of feedback and response comprising all ecosystems, and in fact all living beings on earth: the planetary organism named Gaia proposed by James Lovelock and Lynn Margulis.[14] As they and others have conclusively demonstrated, the mixture of gasses in the atmosphere is far from what one would expect, if geological and chemical processes were running to their normal conclusions. Our planet's air is, instead, a product of living processes: virtually all the earth's oxygen was put into the air, and is now kept there, by biological action. It is a sort of soil aloft—just as generations of plants and organisms slowly build up a soil medium that is conducive to life, so they have also, over millennia, built up and maintained an amenable atmosphere.

What keeps this atmosphere so reliable, breathable, optimal for life? Some vast network of cybernetic interactions. "All of the compartments of the Earth's surface are kept at a steady state, far removed from the expectations of chemistry, through the expenditure of energy by the biosphere."[15] According to systems and information theory, disorder and entropy are inevitable, unless opposed by energy expended to maintain order and organization. The steady state occurs when a system achieves the ability to balance its losses with its gains. The system monitors itself, and when a component reaches a critical state, acts to restore it. Lovelock describes one of the most important of these processes:

> What evidence have we that oxygen is regulated? Certainly for several hundreds of millions of years it can not have been more than a few percent less than now, or the larger animals and flying insects could not have lived. My colleague Andrew Watson has demonstrated in some elegant experiments that it can never have been more than 4% greater than now, and probably not even 1% more.[16]

The earth appears to have a repertory of oxygen-balancing control mechanisms. It produces methane (through biological activity) and buries carbon, among other actions which continually maintain atmospheric oxygen against fluctuation.[17] Similar cybernetic regulation can be described for other parts of the atmospheric and planetary system—other atmospheric gasses and the climate itself, which has absorbed and dampened a gradual increase in solar radiation since life began, of between 30 and 50 percent.[18] The net result of these regulatory systems is that life itself creates and controls its own environment, on a global scale.

The cybernetic perspective leads the scientist to ask unusual questions that would ordinarily be avoided as teleological, and therefore unscientific. "What is the function of each gas in the air or of each component in the sea? Outside the context of Gaia, such a question would be taken as circular and illogical."[19] But of course the parts of a true system do have a purpose or function, some good reason for where and what they are. It is the nature of a self-referential system to create what might be called "meaning" for itself by virtue of its own context-rich network. Once again such questions, which are circular, useless, and unscientific when applied to the inert matter studied by physics, become proper and meaningful when applied to open systems in a steady state.

This embracing system links together the living and nonliving world in a network not merely of physical proximity, but of mental connectedness.

> All cybernetic systems are intelligent to the extent that they must give the correct answer to at least one question. If Gaia exists, then she is without doubt intelligent in this limited sense at least.[20]

This is what Professor Lewis Perelman has called the "global mind."[21]

The catch is that in a group organism such as the ecosystem "it is not easy to point to any part of the system which is the sense organ gathering information and influencing corrective action."[22] The *result* makes it look like a cybernetic system. But how can the *process* be made to fit the cybernetic scheme, in the absence of any identifiable receptors, let alone a central processor?

In fact, cybernetic study of large mentalities such as ecosystems must often make do with the sort of study Walter Cannon conducted of the body's homeostatic regulators. Without knowing exactly where, or how, the regulation took place, he nevertheless could show in detail that it did take place. The regulator was, for him, a "black box," just as many of an ecosystem's actual regulators are for contemporary investigators. As one scientist remarks in a passage quoted by Eugene Odum:

> A model may be constructed explicitly in terms of feedback mechanisms, but often the feedback mechanisms within a system are identified after the model is constructed as "emergent properties."[23]

Lovelock acknowledges that "the information needed to establish Gaia's existence as a control system" is still woefully incomplete.[24]

But Bateson does venture a solution in which the ecosystem feedback circuit operates on somewhat different principles from those of smaller organic individuals. He suggests that "the quantities whose *differences* are the informational indicators are at the same time quantities of needed supplies (food, energy, water, sunlight, and so on)." It may be that information is "immanent" (to use Bateson's term) in these quantities, and that the information processing takes place in the system as a whole. Bateson questions—acknowledging it to be as yet a mere hypothesis—"whether these are analogic systems in which *difference* between events in one round of the circuit and events in the next round (as in the steam engine with governor) becomes the crucial factor in the self-corrective process."[25]

Purpose in a New Perspective

Cybernetic mentalities of this sort, as defined by Bateson and the functionalist philosophers, offer a reasonable explanation for one of the paradoxes of biology: the apparent goal-seeking behavior which occurs at so many levels from the cell to the conscious being. But unanswered questions about purpose remain. Chapter v offered two ways in which mental function and purpose seem impossible to examine scientifically: their intangibility (they seem to invoke metaphysical agents called "minds"), and their backwards arrangement of causality (the end state somehow "causes" actions which precede it). Cybernetic definitions not only demystify the goal-seeking process of homeostasis, but also remove the metaphysics by seeing any goal-seeking mind as a product of syntax, hierarchy, and complex patterning. That leaves the other branch of the problem: what kind of causality can be assigned to purpose?

An answer to the riddle of purposeful behavior may be available through the use of communication theory—the *informational* redefinition of cause and effect, with its implicit reliance on *relation* as the basic character of reality. The philosopher Kenneth Sayre has applied the principles of information and cybernetics in a subtle and far-reaching way that interprets the basic mental process of willing and acting as a very advanced form of systematic information processing. Sayre's work will not be the last word on this problem; but it does illustrate how a relational, information-theoretical approach to the system called "mind" can make inroads on very old and very intractable problems.

Sayre begins with the basic concept of information: the reduction of chance or, put another way, of uncertainty. A single event, once it occurs, carries information (i.e., reduced uncertainty or increased negentropy) of a maximum value: it is 100 percent "probable" because it has already happened. A pattern of information also affects the future by limiting the range of future events. The letters *th* occurring in an English sentence limit the probable next events: the letter *e* will follow a certain percentage of the time, the letter *x* almost never. On a more complex level, an ecosystem limits the probable future states it will inhabit, projecting information into the future. Information limits uncertainty and randomness.

If information is increased probability, then an information channel may be defined as any two events that are probabilistically linked. When an eye sees a star, it receives information: the star is in only one point, and not in any of the infinity of other points it might have occupied. The twinkling of a star reminds the viewer that the information channel is not perfect—the earth's atmosphere diffracts the light-beam irregularly, leading to a small degree of uncertainty. No information is free from uncertainty. But it is information nonetheless.

A cause and an effect, looked at this way, can be seen as two ends of an information channel. The star-eye connection was really a stream of photons that struck a patch of retinal cells, triggering a series of neural charges up the optic nerve. It was a series of causes and effects. A cause is an input, which creates an output effect by drastically limiting the probability that anything else than this specific event can follow this specific cause. To achieve this limitation of probability, the cause expends energy, creating an entropic decline at the effect/output end. This is normal thermodynamics—energy running down as events occur.

Where an information channel is perfectly "noiseless," or lacking in extraneous or accidental randomness, the input and output are identical. That is, the output signal is 100 percent determined by the input; anything else has a probability of zero. This is a description of a physical event after it occurs; its probability is maximum, and its energy-flow is in the normal direction.

In terms of purposeful behavior, this communication-theory approach to cause and effect comes up with a startling conclusion. It is known that biological behavior actually *lessens* entropy within the

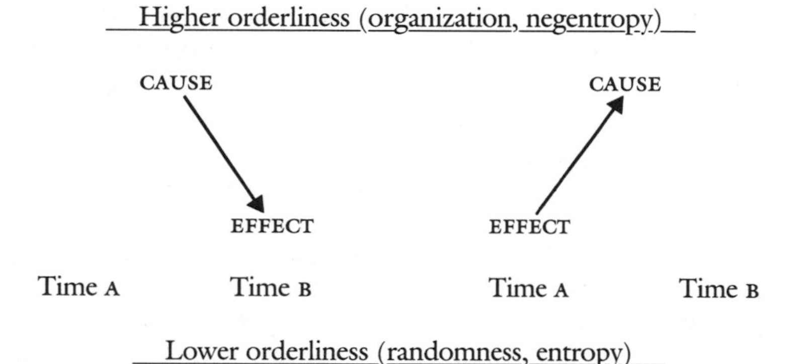

NORMAL THERMODYNAMICS
Entropy increases: the focussed energy at time A is dispersed in movement, friction, heat, etc. Most of it is irretrievably lost. Hence, an "entropic decline" at time B.

BIOLOGICAL THERMODYNAMICS
Entropy decreases: the biological system conceives a goal (the effect to be achieved). Then it achieves it by appropriate action (the cause). The goal makes the cause happen: the effect precedes the cause. Though this uses energy, its net effect is to increase orderliness or negentropy. In entropic terms it runs "uphill" from time A to time B.

local frame of reference. Biological systems, rather than running down, are able to build up over time. Sayre's picture of cause and effect as an information channel allows the interpretation of purpose as exactly that reversed effect-and-cause structure so absurd to ordinary physics.

It works like this. The "direction" of cause-effect is measured in entropy; energy runs down, from higher to lower state, from time A to time B. Biological purpose runs in the opposite direction: effect-cause. But it still makes sense, for it *conserves* the correct distribution of energy. There is still less entropy at the "cause" end than at the "effect" end, as is consistent with ordinary events. (See accompanying diagram.)

In other words, biological systems can be seen as running toward negentropy by reversing the order of cause and effect. They do this by creating so-called "noiseless" information channels that link present mental effects with future states that are their causes. "The noiseless channel . . . enables prior states to be conceived as functions of subsequent states."[26]

There is much more to be said about the problem of purpose, of course, but this sketch suffices to make the point that many important aspects of the biological purpose problem prove tractable, when cybernetically approached. A materialistic universe is governed by strict thermodynamic laws that are contradicted by biological behavior. But the universe defined in terms of its relations, and including a negentropy concept along with its entropy concept, proves able to embrace organic and nonorganic reality with equal ease.

In fact Sayre goes further: he shows that all the higher attributes of life—its evolution to fitter forms, its capacity to learn, and its conscious mentality—are merely advanced versions of well-known characteristics of open, steady-state systems. In other words, he works out, in detail, just how the phenomena of purpose, mind, and consciousness may fit in to a fully natural, nonmystical view of the world.

○ ○ ○

And here lies the final payoff of the ecologistic worldview. The human mind comes home at last. It fully belongs in its world.

The physical superorganism, that large creature of many dispersed parts that was so imaginatively striking in the works of Frederic Clements, and that is still a common metaphor, here appears in a new form. The physical connectedness of the living world is now matched by a mental connectedness.

Thus the ideas of ecologism make a full circle. The wide scope of the superorganism joins together all the physical beings on the planet in complex and surprising interconnectedness. The detailed interactions of the ecosystem make a related point on a narrower scale; complex balances between separate parts make all healthy, make all functional parts of a whole. On a narrower scale still, cooperative interactions often prove more durable and beneficial than competitive ones, as illustrated in the twosomes and threesomes of symbiosis. And the homeostasis observable in the individual's own body further illustrates the kind of complex self-regulation that sets the pattern for all of nature.

But the attempt to actually *understand* homeostasis leads again back to the whole of nature. For the cybernetic concepts of information processing, feedback, and self-regulation suggest natural interconnectedness in an invisible realm: the realm where data is made and used by complex systems and their languages, some of them human, many of them not. And these language-using systems, these natural minds, appear to be linked together in the many layers of nature from the local to the global. The mind of nature is a global, mental superorganism.

This is ecologism's answer to the alienation and self-exile of the human species. It, and all of its thoughts and civilizations and beliefs, truly belong here on the planet. In us the planet thinks some of its most interesting thoughts. And when we contemplate the natural world in which we live, we are contemplating ourselves.

Conclusion: Minds and Their Ideas

This is not to say there is no trouble in the mind of nature, the ecological world. Clearly there is. "Ecology" as a public concern was born of human-caused crisis.

That a thing is natural does not mean it will necessarily survive. Minds, like bodies, may prove badly suited to the world. Their fitness, like the fitness of any other biological system, is continually tested by nature. Extinctions occur.

It is humanity's ideas that will finally decide whether it can pass nature's test—survival. This is the only kind of truth or falsity that it recognizes. Do humankind's ideas lead it to destroy or to create? Do its mental programs cherish life, or disdain it? Does the earth-system, the biosphere of people-animals-plants-weather-etc., prove a steady one, or an unstable one? In the big picture of countless eons and limitless space, of planets and species perhaps in the billions, the temporary blighting of one biosphere is not at all "unnatural." As Paul Colinvaux says, soon it will replenish itself. Extinction of one species, even a particularly interesting and gifted one, is not merely natural but, in fact, inevitable. Though the ecological crisis is real and pressing, its outcome is merely a choice of natural outcomes. So the only question is: will men and women go out by madness and suicide, allowing their ideas to destroy the natural mind/body, or will they learn to cherish the life they are part of and live it out to whatever destiny they may find?

The disruption of the biosphere by a confused human species is our problem—a big one to us, though peanuts in the grand scale. If the biosphere is a global mind, then it follows that this disruption is also a confusion, a disordering of its information circuits—an exact parallel to the human confusion. The two are, in effect, one, since both are linked in a single system. Healthy comprehension of the meaning of nature, of the inescapable network of natural interdependence—these ideas would result in a sane reaction to the problems of worldwide pollution, waste of resources, overpopulation, species extinction, and so forth. But as long as humans persist in their delusion, their crazy notion that they can destroy nature without destroying themselves, then both humans and nature will find their information systems increasingly disjunct, broken, falling back into simpler and simpler, ever more depleted and impoverished forms. "Ecocrisis is global Mind being driven insane by the epistemological errors of human consciousness."[27]

The human and natural world can no longer afford the luxury of selfish and narrow ideas. For it turns out that ideas—values, beliefs, definitions of ourselves and nature—are part of the ecosystem. And the sneering, earth-bestriding human self-image has proven to be a toxic waste.

CHAPTER VII

Ecological Ethics

Every act within the web of life has consequences. In a world in which "everything is connected," ecological awareness leads very quickly to practical questions of action or inaction, to choices that can be neither escaped nor, often, easily weighed.

This is the work of ethics: finding an inner guide, a secure sense of balance and timing, action and stillness. Ethical reflection is like the practice that a dancer must endure before the dance: the hard work that internalizes its form and spirit. A sense of right action emerges by attentive study of the limits and potentials of the body, and the demands of the pattern to be danced.

Right thinking precedes right action. This seems like a truism. But it applies to ecological ethics in a particularly immediate sense; for good or ill, ethics forms an integral part of the natural pattern, because according to cybernetic ecology, human thought is not a phenomenon apart from nature. Individual thinking is simply an unusually intense form of the information processing that appears in so many other natural systems as well. Likewise, forms of group thought, which are so important in developing ethics, are in a real sense nature's own expressions. Language, tradition, ideology, and social structure occupy a higher hierarchical level than individual thought, but are still natural extensions of the principles elsewhere at work in homeostasis, evolution, and the ecosystem.

Hence, to correct errors in thinking or in ideology is not merely to accomplish the logically necessary first step toward correcting behavior. Thoughts and ideologies *are* behavior. They are no less real than rocks or swings of the arm. They are surprisingly influential, and potentially deadly.

Along these lines, Aldo Leopold has suggested that perhaps ethics are a peculiarly human form of group instinct, developing on the level of community ideas. Animal instinct is genetic survival information, automatically transmitted. Ethics, then, would be survival information encoded in our group consciousness—the values and means we teach our children and allow our neighbors and ourselves to pursue. These ideas are as real a part of our world as are the invisible rules that govern the shark and the pilot-fish, or the coevolution that creates an ecosystem. These are all codes that define behavior and limit individual action that would damage the whole. They are the very substance of biological communities. In the human case the codes are not genetic patterns but verbal institutions. But their function is the same.

The community value of limits on murder and incest has long been obvious. Now equally obvious is the value of limits on pollution, growth, consumption, and the like. What Leopold called the "land ethic" extends the meaning of "community" to the entire biosphere: mutual wellbeing is at stake in the environmental realm to which we belong, just as it is in the social.[1]

Van Rensselaer Potter has called ecological ethics "knowledge of how to use knowledge"[2]—a valuable definition that suggests the recursiveness and self-referentiality of ethics. Ethics goes beyond mere technical know-how into an awareness of *whether* and *why*. It replaces the simple linear circuit of problem/solution with layerings of self-checking circuits—an awareness of wider repercussions, a hierarchy of immediate ends and more embracing purposes. Ethics is the wisdom that places human action in its full context. Since individuals seldom can see this full perspective, ethics grows communally, and takes on communal enforcement. The challenge, now, is to recognize the standing of the nonhuman world as part of our community, whose fate is entwined with ours.

A system of ethics, then, may well be seen as an essential regulating device operating at the level of the community. Its purpose is to guide behavior in a way that benefits the individual and the community. It is a natural phenomenon, not in principle different from other forms of systematic regulation. An accurately made ethic is a sort of flyball governor on the complex whirl of human action, limiting where there is excess and urging where there is deficiency.

Ill-regulated human thoughts, ideologies, and values have reached a crisis stage, where they will either stabilize around healthy, biologically valid norms, or destroy themselves and the biological community. A badly made ethical governor can only end in ruin.

Survival: The Ethical Basis

An ethical system must do two things. It must envision some particular goodness or value, which it seeks to maximize. And it must work out ways of doing so that are internally consistent, yet workable in the real world. The particular problem for ecologism is to find a basis for ethics in nature itself. Is there an intrinsic value in nature or in the life process?

For ecological ethics, there must be, or the system fails. Ecologism rests on the notion that the human and the natural do fully coincide. And so the basis of value must be literal and sound, truthful to the facts of natural life as science relates them.

A surprisingly large number of ecological thinkers over the last twenty years (and more) have worked on the problem of finding an intrinsic value in biology on which to build a realistic ethic of life on earth. Their near-unanimous conclusion is that the sciences of life do indeed enable us to find—not invent—a basis for ethical value in the nature of life itself. That basic value is *survival.*

Survival is the master concept of all life forms. This book has presented many examples in which organic systems, on various hierarchic levels, act to preserve their patterns of organization and to project them into the future. Growth and homeostasis, natural selection and evolution, succession and climax, all are organic schemes for survival.

Survival is, in short, the intrinsic purpose of life. And because of this, it is the starting concept of most systems of ecological ethics.

Robert Disch, for instance, subtitles his collection of essays "Values for Survival"; Potter advocates "Bioethics: the Science of Survival"; Ian McHarg comments that all the elements of nature's "intrinsic value system" must pass the "final test of survival"; and so on.[3]

The crisis atmosphere of much ecological writing suggests a kind of rough logic for this emphasis on survival, of course. The unfortunate fact is that a distinctly alarmist shrillness pervaded the environmental movement in its early phases. The popular meaning of "survival" has often been merely for humankind to overcome the ecocatastrophes which seem most imminent.

But behind this rather short-sighted and politicized notion stands the more genuine biological perspective on survival: the twin considerations of the organism's fitness and the environment's health. Both must be continually upheld for "survival" to have any real biological meaning. Survival must be not merely "now" or "where I live" or "for my family," but into future generations, and throughout the ecosphere, and for all those other organisms on whose lives ours depend. The ecological perspective is long and wide—it concerns all of us, here, forever.

It is this broader vision of organism and environment that opens ecological ethics to tremendous development and refinement. Without this expansion, the "survival ethic" does not sound any different from the Darwinian ethic—that horribly stunted notion of kill or be killed. Social Darwinism focused on questions of immediate individual survival, or at most group survival against rival groups. Ecological ethics goes far beyond this by its characteristically enlarged scope of reference. It places the individual within his or her many contexts, and seeks well-being in the most comprehensive sense.

Thus, what survival means in an ecological sense is substantially different than its bare denotation of, say, the momentary not-dying of an individual. It applies simultaneously to individuals, populations, communities, and ecosystems; and it applies simultaneously to the present and to the future. No species has "survived" if its children are left to die in a degraded environment.

Thus defined, "survival" clearly implies a whole range of further values. In working out these further values and the kinds of acts that promote or hinder them, ethicists create a tree-like structure, fanning out in branching ramifications from the root and trunk of the survival principle.

The familiar ecologistic concepts reappear in these ethical formulations. Cooperation and symbiosis, balance, cybernetics, holism—all these, along with their metaphors and myths, carry powerful ethical implications and, occasionally, contradictions. But all of them encourage us to see ourselves as connected to every part of the living world, and to act accordingly.

Values of the Ecosystem: Stability, Diversity, Interdependence

The most pervasive ecological value derived from survival is that of stability: the steady state or dynamic equilibrium. This most striking feature of the ecosystem also appears in the body (as Cannon's homeostasis) and in other kinds of organic structures. But as a model of how a complex multitude of seemingly self-interested individuals can achieve a sturdy survival arrangement, the ecosystem is by far the most telling example of biological stability.

The importance of stability as a biological imperative thus narrows the general goal of survival down to a more specific aim. The dynamic equilibrium of the ecosystem shows us how a diversity of life can sustain itself by cooperation, interdependence, and efficiency (nutrient use and re-use, recycling). To achieve stability is to successfully manage not only the physical conditions that complicate life, but also time itself, insofar as future wellbeing is protected. Of course, this stability depends to some extent upon a reliable physical milieu—predictable climate, infrequent catastrophe, etc. But it also represents an orderliness that sustains itself against considerable challenges of time and chance.

Not the least impressive aspect of this steady state is that it achieves a kind of limitlessness, *within a finite arena.* By its canny dovetailing of roles or niches, by its recycling of nutrients and materials, by its self-limitation in numbers, and by other regulatory feedbacks, the ecosystem creates a new dimension to absorb the pressure to change: the dimension of internal complexity. The ecological vision bluntly recognizes the limits of the finite earth. While the usual directions of space and time are closed and limited for mortal beings, there is apparently no limit to the number of feedback circuits that can be added to a system, no limit to the layering of living pattern upon pattern. Life can limitlessly ramify its own internal meanings, the grammar of its relationships. And while the availability of energy must put some maximum on the number of niches that can be invented or filled, there is no limit to the invention of new cooperative or interactive gambits. Within the syntaxes of living systems, there is room for perhaps an endless deepening of relation and meaning. The ethic of stability does not mean stagnation. It means directing growth and change in nondestructive ways, generated within the standing pattern that supports life.

The most familiar application of the stability ethic is to population and economic growth. The usual formulation combines a perceived threat (crowding, economic privation, decline of quality of life) with an ecological solution. As the famous volume *The Limits to Growth* puts it:

> If the present growth trends in world population, industrialization, pollution, food production, and resource depletion continue unchanged, the limits to growth on this planet will be reached sometime within the next one hundred years. The most probable result will be a rather sudden and uncontrollable decline in both population and industrial capacity.
>
> It is possible to alter these growth trends and to establish a condition of ecological and economic stability that is sustainable far into the future. The state of global equilibrium could be designed so that the basic material needs of each person on earth are satisfied and each person has an equal opportunity to realize his individual human potential.[4]

This is an almost utopian vision. But this group of scientists and economists is able to present it as an actual possibility, because it is patterned after a literal and real model already existing: the ecosystem in steady-state climax.

This vision of a human steady state is central to much of ecological ethics. It implies not only that populations and economies could be restructured for equilibrium rather than growth; it also promises that the continuing degradation of the ecosphere can be halted or drastically slowed. A steady-state use of the forests could halt destructive clear-cutting and replace short-term profit motives with sustainable forest practices. Mineral and oil resources, while impossible to put on a true equilibrium basis, could be used sparingly and more carefully saved for humanity's "long haul." A steady-state style of agriculture could incorporate soil protection with adequate food production, and reduce or eliminate petrochemical pesticides and fertilizers. The continuing encroachment of civilization upon wild and intact ecosystems could be controlled almost absolutely. In short, most of the major threats to the environment that sustains us could be corrected.

These ecological goals would express the values of the ecosystem: long term survival in a steady state, or as close to it as possible. One obvious practical means to achieve these ends would be population reduction: the earth will carry fewer people much more lightly, and with a much wider margin of tolerance. Another, equally important, means, however, would be the development of ecological values not only on the societal level, but on the level of the individual. With the technical means now at humanity's disposal, even a small number can do great damage. The ethical consciousness of earth's human population must therefore be as ecologically well regulated as the size of earth's human population.

On the personal level, the ethic of stability encourages people to conceive of their needs in finite terms, and to learn the satisfaction of "enough." The emphasis on recycling and efficiency are typical choices favored by ecologial ethics. The natural outcome of such values is to focus on quality, rather than quantity. And quality is liable to be measured in fairly encompassing ways. The satisfaction of bodily needs is only the beginning point of measurement. Humans and human groups also rely heavily on intangibles—spiritual, emotional, intellectual, and social qualities. Unless these are well tended also, stability will be very temporary. The ethic of "enough" limits the pursuit of material goods; but it stimulates the pursuit of intangible goods like social relatedness, inward serenity, and personal creativity. As in the ecosystem, a limit is a necessary condition which encourages a deepening in other dimensions.

The contrast to contemporary consumer values could hardly be greater. Advertising and other social influences constantly whet and sharpen our appetites, creating a climate in which only the continual round of desiring and getting and desiring yet more has any status. These self-defeating values embody the paradox that nags all growth-oriented or positive feedback systems. Whatever is achieved instantly becomes inadequate when measured against the ethic of continual consumption. Satisfaction only creates dissatisfaction, in an accelerating cycle. "More" is an unrealizable goal.

Wendell Berry has expressed this contrast of values in cogent terms. Materialist consumerism embodies a static and dead notion of value. Value is "wealth," measured in lifeless tokens to be hoarded or spent. In economics, in politics, in agriculture, and in nature, the consumerist human treats living systems in accord with this lifeless theory of value, as counters to be manipulated and discarded. Modern agribusiness extracts value from the soil exactly the way miners extract minerals from the earth, without reference to the future or to the living character of the soil and crop system. The lifeless Newtonian universe could hardly be better embodied than in these materialist approaches which "use living things as if they were machines."[5]

Ecological ethics offers an alternative conception of value based on its essentially dynamic and living idea of reality. It is a contrast with roots in the difference between a universe of dead atoms and a universe of process and relation.

> We must now ask ourselves if there is not, after all, an absolute good by which we must measure ourselves and for which we must work. That absolute good, I think, is health—not in the merely hygienic

> sense of personal health, but the health, the wholeness, finally the holiness, of Creation, of which our personal health is only a share.[6]

Health is, of course, not a possession or a "thing" at all; it is a process, "a pattern to be preserved." The best one can do is *maintain* health. It exists solely in the present. It requires a kind of ongoing cultivation, like a garden. There is no way to bank any of it, or to have any more than one's share. Only by loving attention to the present reality of the system can it be maintained.

These related expressions of the stability ethic in terms of natural health, sociopolitical equilibrium, and personal values together provide a fairly comprehensive set of guidelines for further ethical action. They clearly advocate styles of living that are balanced and proportionate, that recognize the golden mean in satisfying human needs, and that tend toward nurture rather than adventure, conciliation rather than conquest.

The stability ethic has applications both for the individual and for human societies. We have seen that war has been justified on Darwinian grounds as a fully natural competition whose beneficial result is survival of the fittest. The ecological contrast is total—as is perhaps to be expected in an era when the very meaning of war has utterly changed. More than one ecological writer has looked at the nuclear confrontation as a biological survival issue whose outcome can only be expressed in terms of cooperation or death. "Survival through dynamic equilibrium" between the nations is the hopeful model offered by the ecosystem.[7] According to some political scientists,

> The ecological way of seeing and comprehending envisages international politics as a *system of relationships* among [interdependent earth-related] communities.[8]

Thus on all levels, from the individual to the group and from the local to the international, survival requires maintenance, health, and stability. Ecological ethics chooses modes of living and acting which do not substitute temporary individual gain for overall stability. For the health of the whole is primary.

○ ○ ○

Not surprisingly, the ethic of stability and permanence is often associated with a related value: diversity.

The importance of a rich variety of different organisms and ecosystems may be presented, in typical ecological ethics, in several ways. Diversity may be seen as intrinsically good, a fundamental value choice that is virtually self-evident. No human wishes to live in a sterile monotony. And in fact, earthly evolution as a whole has travelled in the direction of greater and greater complexity. While evolution is never predetermined—an outcome of decline, extinction, and lifelessness is also possible—nevertheless, on the whole, so intense has this complexification been that living processes have reformed the earth's air, sea, and land until they are almost alive themselves, crammed with organic materials, stabilized around biologically determined norms, and capable of supporting earth's stupendous variety of life.

The value of diversity may also be more directly derived from the stability ethic. There seems to be, as we have seen, a reciprocal relation between ecological stability and diversity: stability in general promotes diversity; and in some cases, diversity in turn contributes to stability. The slow succession of ecosystems that builds toward the climax steady state represents a dramatic increase in numbers of species and interactions. The simple quantity-oriented pioneer (or "weed") stages give way to increasingly subtle interactions among more and more kinds of creatures. Again, this is not an absolute rule; some climax states represent a slight decrease in diversity compared to immediately preceding states, though still great increases over the earliest stages. In general, the *quality* of the ecosystem's life improves, in the sense that its complexity, diversity, and stability increase.

Ian McHarg identifies diversity as the best measure of natural health. Diversity and complexity can also be denominated as *order;* within a steady-state ecosystem, such order also signifies an elaborate equilibrium. This high level of orderliness is found in patterns of close interaction and layering: namely symbiosis, consciousness, and negentropy. According to McHarg, these features tell us that the natural system is functioning. Its circuits are intact, and its powers are organized in stable ways that maximize the numbers and kinds of living beings that can survive.[9] On any level, from the most local to the most global, the health of the living environment is often reflected in the diversity it supports. When our actions tend to simplify nature, we may be sure they are unhealthy for the ecosystem as a whole.

From this description of the thriving and involuted networks of life, it is clear that stability and diversity must imply a further ethical norm: *interdependence.* This is a subtle ethic that, interestingly, blends maximum variety with maximum cooperation and mutuality. For ecological interdependence occurs not among identical, machine-stamped parts, but among unique and widely varying roles and kinds. Yet these are melded into unity.

This combination of unity and diversity forms a practical ethic of wide application. Bruce Allsopp, for instance, interprets it simply as "toleration," and applies it in the personal, social, and political arenas. If one regards "the world as an organism in which all the living parts contribute to the whole," then racial, religious, political, and national distinctions cannot lead to strife.[10] The ecosystem teaches us to see even our opponents as performing a necessary role; and beyond that, to see a common good which binds opposites together.

Recognition of this unity in variety leads to an acceptance, even a love, of diversity, oddity, peculiarity, and uniqueness. For this variety is the very substance of living systems. To love what is very different from oneself is perhaps the ultimate test of emotional and spiritual maturity. It is a direction of moral growth strongly encouraged by ecologistic thinking.

One of the most striking results of this ecological value is an intense opposition to the uniformity and sterility common in modern urban and industrial life. The tendency of the biosphere is toward increase in order, but this means the opposite of uniformity. Only simple, "weed"-stage ecosystems are relatively uniform. Organic negentropy implies the endless variations of life we see around us in thousands of interwoven species.

The machine metaphor of our dominant social values grinds rather inexorably toward ever greater regimentation and simplicity in products, social institutions, and (most dangerous of all) thought. But envisioning society as an ecosystem, with many roles and niches, each essential to the organic health of the whole, offers a strong antidote. Ecological values move toward more social choices, not fewer; toward decentralization of power and authority; toward greater individuality of style and expression.

Environmentalist resistance to centralized power generation, for instance, is well known. When solar and other renewable sources became a practical possibility, one of the hoped-for benefits was eventual public dispersal of a very powerful concentration of capital and influence—the energy companies and utilities. Literal "power to the people" (in the form of rooftop solar collectors and such) would destroy a too-great public uniformity, and open up personal choice and control over basic creaturely needs. The electrical power grid, with its few, vastly expensive and heavily guarded centers, and its radiating lines reaching into each citizen's home, might be taken as a pretty fair visualization of just what "big power" means socially. It means decisions made at the top of the social pyramid, and individuals expected to conform. Oil and nuclear power work the same way. But dispersed, individually operated solar collectors would lead to a variety of solutions, a maximum of choice, and a minimum of regimentation.

Such a diffusion of power sources would also create a much more stable and secure economy. According to Amory and L. Hunter Lovins, these "dispersed, diverse, and redundant systems" would be virtually immune to foreign attack and domestic breakdown. Neither blackout, nor terrorist sabotage, nor blackmail by international cartel could touch them.[11]

This combination of tangible and intangible benefits is typical of ecological ethics. The *right* solution addresses problems as complex questions of many interlocking levels. It will not only solve the group problem, but feed the individual spirit as well. A holistic solution acknowledges the importance of all levels, and searches for the most flexible and multifaceted approach. It understands that a certain ebullience and occasional sloppiness are the price of variety. It accepts them as part of the process.

Hence an ecological ethic supports choice and variety in most issues. It prefers old neighborhoods to new tracts, relishes cranky and idiosyncratic beliefs over mass orthodoxy. It hopes food will continue to come in irregular forms— not always packaged for ease of national distribution or bio-engineered for the convenience of machine harvesters. It supports a range of spiritual and sexual options. It believes that a person ought not to be forced into an unnatural frame of behavior or belief. Whatever loss of antlike uniformity or efficiency this variety causes is more than repaid by the increase in vigor, diversity, and stability of the whole.

No ethic can wholly erase the tension between individual ends and group ends. There is always bound to be some conflict—else there would be no need of ethics at all. But ecological ethics sees a twofold way to minimize the disjunction. First, by allowing maximal latitude and tolerance. And second, by shaping that latitude to stay within biological necessity. When ethical limits are felt not to be arbitrary, but to be a part of the very life process itself, then a high degree of voluntary accord is possible. A serious commitment to life-as-it-really-is draws the ecological diversity into a substantial unity.

Three Contradictions

It is perhaps inevitable that ecological thinking should sometimes appear to make contradictory choices. The richness of the ecosystem and the processes of life offer many models. The metaphor of the superorganism dominates ecological thinking in some situations. It emphasizes the oneness of parts, their subservience to the whole. In other situations, visions of interdependence, symbiosis, and cooperation imply latitude of choice and decentralization of control: the arrangements of life made in many small choices and separate *ménages*. Similarly, the reality of change must be acknowledged in any system; but so must the importance of homeostasis and stability. In the ecosystem, both sides of these opposites are present.

But the human must make choices, discover how to balance the apparent opposites of public and private, change and stability. For society does not usually allow nature to take its course, slowly killing off the wrong solutions (or individuals), or allowing whole segments to be wiped out by a hostile or selfish invasion, as may happen in an ecosystem from time to time. Human controls seek to operate at the level of the idea, not at the level of brute outcome. How to define the highest social good, and how to convince individuals to seek it, remain troublesome questions.

How can we combine the values of diversity and interdependence, for instance? With the example of the ecosystem firmly in mind, private action that serves the greater good of the whole does not seem impossible. But in practice, ecological thinking wavers between individualism and communalism.

Though it promotes the health of the whole, ecologism also encourages a particular kind of independence. Since the majority culture is regarded as sick and unnatural, an ecologically sound life will have to disassociate itself from the relentless militarism, conformism, and consumerism all around it. This will look like eccentricity, and may lead to isolation and withdrawal from mass culture. Such a life may also detach itself on the physical level, turning toward self-sufficiency in power and food sources. But this independence from mass culture is also an intensified, personalized *dependence* on natural processes.

On the social level, this ecologistic independence from mass culture will be paired with an increased attention to more immediate communal and personal relations. The relevant social unit is those people in the daily circle of work and living. Responsibility to one's immediate group and local biological/land system characterize this kind of living, for one is in direct, daily contact with both.

The immediacy of this focus on real neighbors and a local patch of earth contrasts vividly with the typical pattern of industrial culture. For most urban dwellers are strangely split between two levels, neither of which allows them direct involvement and responsibility for very many other humans or for any part of the natural world. On the one hand they will cling to an emotional identification with some large, abstract social entity, like the "nation," or a political faction, or even a particular sports team. Their principal connection to these will be through television, which they watch in a silent, semidark room which may or may not contain other people (it matters little). These identifications will arouse great emotion, but little thought or action. Such images are notoriously susceptible to demagoguery, for they have little content except as artificial vehicles for the emotional need to belong. The watcher's hunger is thus pathetically manipulated for profit or power, but never truly fed. The mindless bellowing of the "fan" is the commonest behavior in all three cases. Here conformism is absolute. People will follow the leader to any war or folly, as history shows, if they are sufficiently incited.

In daily life, by contrast, these urbanites will live in almost total detachment from place, weather, season, or neighbors. Here they are virtually independent. Their lives will contain no form of responsibility for the wellbeing of the people around them, or for the natural systems of the area. They hardly know they exist.

The ecologistic approach changes the mixture of dependence and independence. It cuts off the abstract, bodiless images of mass communications, ideology, etc., and replaces them with connectedness to the unmediated realities near at hand. It is communal in one sense, fiercely independent in another. Where it senses a connection to a real social organism, it will probably choose group welfare over the individual will.

Where ecologically minded people are involved with public issues and politics, they will similarly tend to side with communal over private interests, seeing only a spurious individualism in the private acts that profit at the expense of public lands, breathable air, and the like—the sick individualism of irresponsible, unconnected people. Some of the solutions called for by environmentalists—for example, in pollution control—can be branded as authoritarian, top-down social control. Although many political conservatives were conservationists during the early decades of the century, they have found themselves opposing the new environmentalism as an oppressive extension of Federal control over "private" (usually meaning "business") freedom of action. Land use and wildlife protection are other issues on which ecologism usually lines up on the side of communal control, against the unregulated actions of individuals.

Only by keeping in mind the way the variety of ecological metaphors and images is held together by a felt sense of organic unity can these contradictions make sense (which is not the same as claiming that they are necessarily logically consistent). Usually this sense limits diversity and individuality only where these are perceived as threatening the more basic considerations of survival and stability. Where the survival of the whole seems threatened, as it is in issues of extinction and pollution, then the basic ethos of protecting the whole predominates.

○ ○ ○

The ethos of protecting the whole originates in what has been variously called biocentrism (the opposite of anthropocentrism) and biophilia (love of life). These words extend humankind's habitual self-love to the entire living world. This ethical and emotional extension results naturally from the fundamental ecological insight that humans are part of nature, not separate from it or somehow the opposite of it. As Aldo Leopold pointed out, the "land ethic" derives from recognition "that the individual is a member of a community of interdependent parts," a link in the mutually supporting network of nature.[12]

But what happens when this love of the whole notices that humankind itself is the greatest threat to the rest of the living community? This perception triggers a larger form of the individual-versus-group dilemma, pitting the individual human species against the rest of the biosphere. In Chapter II, Loren Eiseley was quoted asking the rhetorical question, "Is man but a planetary disease?" In recent decades, the answer has been an ever more emphatic "Yes." One might well conclude that the cancer ought to be eradicated.

In this way, biophilia is transmuted into misanthropy—a strangely unexpected contradiction to discover in the heart of ecologism. So great may be the love of nature, of wilderness and the nonhuman world, that it calls forth hatred of civilization, even hatred of *Homo sapiens*. The ironic extreme of earth-love may sometimes be human-hate.

This is not as rare an emotion as it may seem. Aside from those numbed by economic self-interest, who does not feel a brief flash of anger and loathing at the sight of a forest reduced to slash, a strip mine, a polluted beach, a dead river, a stinking brown sky? Who does not "side with the bears" (in John Muir's phrase) at least occasionally when comparing the works of nature with the follies of civilization? Just as a satirist, like Jonathan Swift for instance, may begin with an acute love of virtue, but end in a raging hatred of vice, so it is easy to find one's love of nature slowly changing into hatred of nature's apparent enemy, humankind. As we will see in Chapter IX, this conflict of values has not remained merely potential within the ecological worldview: the organization called Earth First! has openly espoused near-violence and even cheered for pestilence and famine—all on behalf of the beleaguered environment. This is the dark side of ecologism, its shadow. In it the ethic of life-love emerges looking strangely like its own opposite.[13]

An effective answer to this painful contradiction may lie in Leopold's key term, community. If humans are part of the natural web, then their legitimate interests, their unique contribution, deserve respect too. Birch and Cobb point out that community does not mean some kind of abstract equality; rather, it means "that we share one another's fate."[14] It certainly does not mean choosing one part of the community over another. To do so is merely to repeat the error which causes the problem to begin with—the mistaken belief that one part is separate from the other. In the community of life on earth, the choices must be made in a context of inclusion: not either nature or people, but nature and people. For the deeper truth is that humankind is nature.

A truism does not always help resolve actual problems, however. Some—much—human behavior threatens the wellbeing of the earth. It must be changed. But the struggle to change anti-life behavior must be undertaken as a loving labor within the commune of all life, not as a commando raid on an enemy camp. Otherwise, the champions of life will discover themselves to be instead the agents of death. They will have become that which they most feared.

○ ○ ○

A third major conflict in ecological thinking breaks out between the ideas of progress and stasis. Because the dominant culture so thoroughly promotes progress, ecologism (like much of this book) often concentrates on refuting it. Few ecological appeals get much beyond images of homeostatic stability and arguments on the impossibility of endless growth.

Yet this emphasis might easily be misleading: ecologism does not seriously plan to freeze time and stop change. What it hopes to do is to achieve substantial equilibrium with the natural world, so that the means of life are not destroyed. Gradual and evolutionary change will always continue—but it must be change within the limits of survival. The conflict, of course, occurs in trying to distinguish good change from bad change, genuine progress from cancerous growth, stasis from stagnation.

The idea of progress touches most directly the question of technology. Given the fact that unconstrained technology, in the form of industrial civilization, has caused a good deal of the ecological destruction around us, it can be seen as the crux of the environmental problem. But the point is not nostalgic yearning to go back to a paleolithic hunter-gatherer society. A fraction of the environmental movement might be simple back-to-nature romantics, but this is not a serious response to the situation.[15] The apple cannot be put back on the tree, and our steps cannot be backwards-taken. Artifice is part of human nature; the question is, How much artifice shall we have? How much technology, and of what kind?

There are really two questions here: one for the industrialized nations, and another for the Third World or so-called "developing" nations (an unfortunate term that assumes these countries must be moving toward the same sorry state as the industrialized countries). We will look at them in sequence.

To state the obvious, industrial technology is complicated, expensive, and polluting. Its complexity must be supported by vast infrastructures of social bureaucracy, subsidiary technologies, and transport. Its cost is visible not only in terms of the massed capital it requires, but also in the large amounts of raw materials and power it uses up. Then the environment is further burdened by having to absorb both the effluents of technological production, and the waste produced by excessive packaging and throw-away product design.

Such technology conflicts with the ecologistic vision because it is not sustainable. The industrialized West consumes a huge proportion of the world's power and materials. In the decades following World War II, when high technology rapidly accelerated, production of manufactured goods in many categories—plastics, synthetic fibers, nonreturnable bottles, fertilizer nitrogen, and so on—shot up at rates in the thousands of percents.[16] Now, as economist Dennis Pirages remarks, "The mast majority of humanity finds itself running harder just to stay in place on an economic treadmill."[17] The subsidy of cheap fossil fuel has begun to be withdrawn, the waste problem has begun to mount, and the glittering technological future begins to look highly questionable.

Though high technology seems to work wonders by providing cheap, easy solutions for its fortunate few customers, in reality those solutions are full of hidden costs. How easy it is to pour a gallon of gas into a power mower and rev it up, or spread out a few pounds of fertilizer and insecticide, and thus effortlessly gain a beautiful green lawn! Yet that gallon of gas is drawn from a finite supply which, though not yet ready to run out, will one day do so. When it does, people will rightly ask whether the suburban lawns were worth all the petroleum and petrochemicals wasted on them—particularly when the same lucky suburbanite may, later that day, pay for the privilege of going to a gym to work up that same sweat he or she just avoided. And one wonders how much gasoline and petrochemical fumes the faithful exerciser will inhale from the smoggy air. The price of over-reliance on technology is depleted resources and a polluted environment —prices whose payment is merely deferred or disguised.

Technology has become divorced from a sane relation to the totality of human life. Jacques Ellul has called technology the "science of means."[18] Technology typically sees only as far as the immediate end, and provides the "best" way to get there. So far, humans have seldom been able to resist this "best" way, because they have seldom been able to place the immediate end within a wider framework of total ends and means.

In his now-classic book *Small Is Beautiful: Economics as if People Mattered,* the economist E.F. Schumacher has written eloquently on the relation of ends to means, the question of what kind of technology can get us the kind of life we really want. He points out that

> Technology recognises no self-limiting principle—in terms, for instance, of size, speed, or violence. It therefore does not possess the virtues of being self-balancing, self-adjusting, and self-cleaning. In the subtle system of nature, technology, and in particular the super-technology of the modern world, acts like a foreign body, and there are numerous signs of rejection.[19]

High technology violates the ideal of the steady state. It is a runaway train, a one-way ticket. Whoever comes after us will have to live within a used-up landscape of useless fragments, and will have to create something simpler, something more in harmony with the world's limits.

There are yet deeper problems with the culture of high technology. Even if it were sustainable, would it be desirable? For every activity, says Schumacher, there is an appropriate scale. "What scale is appropriate? It depends on what we are trying to do."

> Take the question of size of a city. While one cannot judge these things with precision, I think it is fairly safe to say that the upper limit of what is desirable is probably something of the order of half a million inhabitants. It is quite clear that above such a size nothing is added to the virtue of a city. . . . The finest cities in history have been very small by twentieth-century standards. The instruments and institutions of city culture depend, no doubt, on a certain accumulation of wealth. But how much wealth has to be accumulated depends on the type of culture pursued. Philosophy, the arts and religion cost very, very little money. Other types of what claims to be "high culture"—space research or ultra-modern physics—cost a lot of money, but are somewhat removed from the real needs of men.[20]

The size and complexity of our endeavors must be brought into relation with humankind's real needs: for meaningful work, for intimate contact with others, for enough security and enough challenge, for air, water, and food, for rootedness in a place, and for some way to touch the mystery of nature. Against these actual needs, technological society runs automatically toward ever greater complexity and size. It does so without any ability to question whether lives lost in hugeness and tangled in abstract social mazes are lives which humans want to live. So the ecological question of how much technology nature can carry is mated to an exactly parallel psychological question of how much technology human nature can stand. The answer to both is, not coincidentally: a limited amount.

What is needed is "technology with a human face" (Schumacher's phrase)—a recognition that the giantism of our corporate and economic lives is a deficit upon the spirit and the earth. Paul Goodman has made a very simple, yet penetrating suggestion for humans caught in the whirling acceleration of industrial society:

> Since we are technologically overcommitted, a good general maxim in advanced countries at present is to innovate in order to simplify the technical system, but otherwise to innovate as sparingly as possible.[21]

Innovation to simplify is appropriate technology. A really good tool, human powered, may replace a really ugly, noisy, stinking, expensive, hard-to-keep-running high-tech tool (like a power mower). A really good rooftop solar collector is not only simpler than, but economically, socially, ecologically, and aesthetically better than, a nuclear reactor which will take years to build, employ only the most highly skilled, spew out thermal and radioactive pollution, and soon (after thirty or forty years) have to be decommissioned—to sit for centuries a steel and concrete menace. To say the phrase "appropriate technology" is to ask "appropriate for what?" And then to answer in a way that frames the narrow question of means (technique) within the larger question of ends—how human lives are to be lived. Naturally, only a conscious human mind can do this; left to itself technology ignores the problem and builds itself ever more lifeless mazes for ratlike humans to run in. To connect means to ends, and narrow ends to larger ones, is to construct an ethics to live by.

The ecological movement does not, therefore, pretend that technology is inherently bad or must be eliminated; it merely seeks the right technology for the job, be it simple and elegant or powerful and complex. Shaggy individualists in the north woods might be found toying with advanced solar technology, even while they insist on outdoor plumbing and mule-powered farming. There is room for virtually all levels of technology in a sanely ordered world—though of course one needs a compelling reason to justify the really exotic levels, with their high costs in resources and pollution.

And on this point, differences within ecologism can be seen. When is really high technology justified? Though some still mistrust the computer, many of the environmentally minded (including Schumacher) embrace it as a fine example of dispersed, redundant, highly personalized high technology. Marine environmentalist Jacques Cousteau, to offer another example of appropriate high technology, has created a fancy new ultra-technological wind-ship (the "Alcyon"), on the theory that intelligent application of design can replace unintelligent reliance on fossil fuel. Spaceflight, on the other hand, is an issue that has divided otherwise like-minded people. *Deep Ecology* associates it with a throwaway mentality: we can escape our planet and inhabit the galaxy, so why worry about saving the earth?[22] When *CoEvolution Quarterly* ran an essay about space colonies in 1975, an explosion of reaction followed, both pro and con. Some saw spaceflight as part of humankind's expanding consciousness, a delicious and apt fulfillment of its role as part of Gaia's brains and eyes. Others decried its escapism and reliance on a polluting and rather militarized industrial culture.[23]

The ecologistic vision includes both simple and complex technologies, both progressive change and carefully husbanded stability. The choices between them are occasionally perplexing, but are guided (at least in principle) by an awareness that no end may be narrowly conceived: all purposes must be fitted to the overarching necessity of maintaining appropriate scale and environmental health.

❍ ❍ ❍

How does this wider, wiser perspective on technology view the other question—that of the Third World? It may be very well to try to make those who are already well fed and housed a bit more environmentally sound. But what of the millions who are hungry and unsheltered—don't they await the coming of industrial deliverance?

It is not hard to see that taking the industrialized world as a model for development would only invite these countries to repeat Western mistakes and worsen an already damaged planetary environment. But in addition, Western-style industrial development simply does not answer the real needs of countries with large populations, underemployment, and little money. In fact, mass production is designed for the opposite conditions: "It has arisen in societies which are rich in capital and short of labour and therefore cannot possibly be appropriate for societies short of capital and rich in labour." In India, in Africa, in Puerto Rico—wherever it has been tried—the result of well-meaning technological industrialization in the Third World has frequently ended up far wide of the mark. It sops up huge amounts of capital and creates a few jobs for the highly skilled; but often it cannot otherwise help the region because it requires highly specific and refined raw materials, and because the end product is irrelevant to the lives of the local inhabitants—too complex and too expensive.[24]

The consideration that matters in the Third World is not to maximize output efficiency and quality, but rather to maximize human participation, product affordability, and product relevance. The way to do that, as stated in Schumacher's 1965 United Nations proposal, is to promote "intermediate technology" that will create "literally millions of workplaces."[25]

> The system of *production by the masses* mobilises the priceless resources which are possessed by all human beings, their clever brains and skilful hands, *and supports them with first class tools.* The technology of *mass production* is inherently violent, ecologically damaging, self-defeating in terms of non-renewable resources, and stultifying for the human person. The technology of *production by the masses,* making use of the best of modern knowledge and experience, is conducive to decentralisation, compatible with the laws

> of ecology, gentle in its use of scarce resources, and designed to serve the human person instead of making him the servant of machines.[26]

This is not primitivism or romantic nostalgia for the supposedly simpler past. This is a call to use the best efforts of the engineer and the scientist to create better ways of achieving simple goals simply—in other ways, to create better tools.

Producing this kind of technology would not be charity work by the rich nations. For it turns out to be very similar to the kind of technology we need at home as well: tools that do not require bureaucracies of managers and armies of resource miners and transporters; tools that do not burden the earth with pollution; tools that do not bind the life of the user in needless complication. Such would be the right kind of progress—progress in spirit and in cleverness, progress that aims to live in equilibrium with the limitations of the mother planet.

❍ ❍ ❍

Ecologism includes potentially conflicting values of both change and stability, individuality and collectivism, biophilia and misanthropy. Other conflicts could be explored: the primitivist strain which tugs against the scientific basis of this worldview; or the belief in the frailty of "our precarious habitat,"[27] which is undercut by a deeper sense of the permanence of nature. These and other contradictions stem from the rich variety of concepts and metaphors offered by the ecological worldview. The ethical choices of ecologism move, as human thought is wont to do, through a tortuous labyrinth that is only partly logical. These choices take their cues from nature, as understood through ideas like cooperation and stability, and as pictured in exemplary models like symbiosis and the climax ecosystem. As we have seen, these ideas and pictures do not always carry a consistent message.

Carolyn Merchant has pointed out how images of nature like these influence cultural behavior, despite a scientific tradition that forbids mixing the ethical and the scientific.

> It is important to recognize the normative import of descriptive statements about nature. Contemporary philosophers of language have critically reassessed the earlier positivist distinction between the "is" of science and the "ought" of society, arguing that descriptions and norms are not opposed to one another . . . but are contained within each other. Descriptive statements about the world can presuppose the normative; they are then ethic-laden.[28]

Since some of the images of ecologism are contradictory, however, it appears that a deeper, less conscious principle must govern the choice between them. It is, apparently, a respect and love for the ecological whole that operates as a guiding principle, reducing the confusion between the various images and aims of the ecological worldview. In operation it may seem almost like an aesthetic rather than a rational ethic, for it can spring from the whole person, not just the intellect. An ecologically centered and grounded mind will respond with unconscious sureness to many of the difficult ethical choices. This is not because it is irrational, but often because its reason is integrated into a sense of its total biological life. And choices that run counter to this totality will often be marked with an obvious ethical ugliness.

The Best Possible World?

These reflections lead to the awareness that some value choice precedes the value of survival. Undergirding the ethic of survival must be the notion that *life is good.* If it were not, then we could not regard its survival as a valid ethic.

How good? The ecological answer seems to be: very good indeed—better than anything else we could imagine. This fundamental tenet is at one time a belief, a feeling, and (as we will see) seemingly a proven fact. Some formulations of ecological ethics rely directly on this base, making survival and its various requirements subsidiary. As we have seen, most ecological ethics at least unconsciously rely on some sense of life-love as a partial guide in choosing among the various ecological images and models of action. Ultimately, it is clear that ecologism leads to an absolute love of being, in the sense of actual life, living physically in the real world. This love tends to be impersonal: it is focussed not on the self but on the whole network of living selves. And this love cannot be compromised by any rival value whatever.

Theologians Charles Birch and John B. Cobb have formulated their vision of the world's value in precisely these terms, in their book *The Liberation of Life: From the Cell to the Community* (1981). Their foundational principle, the goodness of life, leads them to a well-distilled ethic: "maximizing richness of experience." If life is good, then more of life is better. But "more" must be understood qualitatively. It does not simply mean the greatest possible number of kilograms of flesh crammed onto the planet's surface. Instead it would occur in the greatest variety ("richness") of living experience, including that of the nonhuman world. Human consciousness is not the only experience to be measured by this yardstick. The lives of cells, plankton, plants, animals, and the various biotic communities have ethical standing as well.[29]

Maximizing value, so conceived, means most of what other ecological ethics mean: survival, stability, diversity, interdependence. Incontinent population growth among humans, for instance, is singled out for criticism, here as in other ecological ethics. The values of larger life-units are stressed. An ecosystem is real enough to represent an important level of biological experience. Responsible ethics means, in the many choices we face, accepting limits on personal activity, in order to encourage the greatest possible quality of life on our planet—for ourselves, for our children, and for our other fellow beings. The ethic of Birch and Cobb, then, is not really dissimilar to the survival-based ethic more commonly encountered. It simply begins from the more basic value: the goodness of life and our love for it.

❍ ❍ ❍

An ethic that ultimately depends not on love of God or faith in revelation but on the goodness of life must be ready to answer the hard questions humans have about pain and death. Life certainly exists, and works by certain processes. But is it really *good?*

The familiar pattern of ecological thinking, as we have seen in Birch and Cobb as well as many other examples, is to look always at the larger picture. To find out the function, or purpose, of a puzzling or painful fact, one must refer to its place in the larger scheme. The "ugliness" of the blowfly or deepsea fish, the "horrors" of predation and parasitism so dwelt upon by Darwinists, the pains and fears of bodily life—these challenge our moral sense, but do not prove to be meaningless in the economy of life. As Alan Watts says, in his book (significantly titled *The Joyous Cosmology*):

> For in this world nothing is wrong, nothing is even stupid. The sense of wrong is simply failure to see where something fits into a pattern. . . .[30]

The Western tradition most like this ecological optimism is that of eighteenth century Rationalism. One readily recalls Dr. Pangloss, Voltaire's bitter satire on the notion that this could be "the best of all possible worlds." But stronger than this undercurrent of doubt was the conviction of deists and rationalists that, in Alexander Pope's well-known formulation:

> All Nature is but Art, unknown to thee;
> All Chance, Direction, which thou canst not see;
> All Discord, Harmony, not understood;
> All partial Evil, universal Good;
> And, spite of Pride, in erring Reason's spite,
> One truth is clear, 'Whatever is, is RIGHT.'[31]

Ecologism rediscovers this kind of confidence and assurance. Naturally, its basis is radically different. Ecological thinking relies on scientific reasoning, not reasonable faith. But its conclusions can be remarkably similar to those of rational deism. Both worldviews see the universe as an orderly place, in which humankind fully belongs, and in which what Pope called the "universal system" contains full justification for any particular fact within it.

How far can ecologism push such a sweeping affirmation? Since it is limited to the natural world, probably less far than Pope could.

There are, after all, different kinds of ugliness and pain to "justify." Some is naturally inherent in life: mortal beings are limited, and so must suffer limitations on their power, desires, and (ultimately) lives. They are physical, and so must accept the accidents of the flesh. For these pains, the ecological worldview offers a profound acceptance, even love. The limits are a necessary part of an astonishingly beautiful whole.

But humans are also social, intellectual, and moral beings. And plainly, in these arenas some of the world's pain is generated gratuitously—not a necessary part of life, but in fact a result of aberrations from it. An ecological approach to living would perhaps minimize these distortions, by creating lives not in fundamental disharmony with the biological realities.[32] This removal of distortions leaves open the possibility of spiritual growth, serenity, depth of inwardness. It removes unnecessary impediments. But how far one goes, and by what path, are questions not really within the ecological purview. Some writers offer suggestions about kinds of spirituality consistent with ecological thinking—usually nonauthoritarian and nonaggressive forms; often meditative; always tending to build an inner life of humility and perceptiveness toward the outer life of nature. But these suggestions take the reader outside the limits of the worldview proper. The mysteries of the inward life can be illuminated, but not answered, from the life of nature.

Thus the ecological cosmos offers a fairly complete natural theodicy—a systematic defense of the goodness of life and the forces that bring it forth. In traditional theology, such a defense ends by affirming God's goodness. The ecological theodicy ends in Gregory Bateson's "sacred unity of the biosphere," an earth-oriented affirmation of the innate goodness of life. "Whatever the ups and downs of detail within our limited experience, the larger whole is primarily beautiful."[33]

The feeling of living in such a universe is a bright contrast to life in the twilight of sin depicted by traditional Christian theology, and further yet from the post-Christian dungeons of existentialist thought. Imprisoned within the isolated human consciousness, both Christian and existentialist could only find the limits of mortal life galling—

and look for deliverance. But deliverance from the ecological cosmos is never thought of, for it is not merely the stopping-place, but the *home* of life. This very universe has brought us forth, and all our powers and limits are but parts of an immense power and limitlessness. We are not pilgrims passing through. We are life itself.

This feeling of the universe's rightness has been worked out in a surprisingly scientific form in Lawrence Henderson's *The Fitness of the Environment.* Though written before the ecological movement (in 1913), it is a volume quoted admiringly by ecologists such as W.C. Allee. It asks the question, "To what extent do the characteristics of matter and energy and the cosmic processes favor the existence" of life?[34] Is our existence, as we have been told, some kind of bizarre cosmic aberration, an improbable fluke that merely disguises the hostility of a cold, huge universe?

Henderson examines the elements crucial to living structure—carbon, hydrogen, oxygen. He observes that the characteristics of these, singly and in combination, favor life far beyond the characteristics of any other elements extant. It is not merely that life as we know it happened to evolve using these materials. Rather—assuming that life requires complexity, regulation, and metabolism—it is that these materials are by a wide margin *uniquely suitable.* In freezing and boiling characteristics, in conductivity, combinability, density, and scores of other measures, none other of the universe's elements could possibly serve so well to foster life.

And these elements, apparently so uniquely fit for life, are the most common in the universe. They are abundant everywhere.

After reviewing the possibilities for life inherent in these elemental materials, Henderson (and with him many ecological thinkers) reaches a startling conclusion. What we see around us is no accident. Life is no fluke, no exception to a larger and grimmer reality. We live in "the best of all possible environments for life."[35] Ecology teaches us to see that the organic processes of life and death are beautiful, lovely, tender, and proper for life; Henderson goes a step further and shows that the inorganic processes are equally so.

The very atoms in our bodies, one and the same with the elements in the stars and solar systems, are laden with potentiality for life. The universe does not merely tolerate life: it encourages it. Life, in some basic way, seems to be virtually implicit in the basic physical structure of the cosmos.

○ ○ ○

Beyond this it does not seem necessary to go. Ecological thinking, for all its variety and its sometimes contradiction, comes to rest in this sense of rightness. We are here because we belong here. Our lives and minds are natural; are, in fact, nature. The places we live—plot of earth, planet, stellar system—suit us. What we do and think is a part of what these places are. Here is, perhaps, the hidden connection to ostensibly alien philosophies from the Orient, which inevitably creep in to ecological discourse. It is this repose, this refusal of anxiety. We belong here.

It is this sense of belonging which ecological ethics seeks to bring fully to our civilization and our way of living. Ultimately, this aim defines a peculiarly circular situation for the human species. If humankind adopts the ethics of belonging to the earth, then it will indeed belong. It will fit in; it will be fit; it will survive. Its ethical codes will prove truly natural, and pass the test of survival. If on the other hand it does not adopt this sense of belonging, then this not-belonging too will "come true": humanity will inevitably pollute, breed, or explode itself out of existence. It will have proven maladapted, failed the test of survival.

The metaphors, images, thoughts, sciences, emotions, and ethics of humans are thus part of the great unfolding drama of evolution. They will finally decide the fate of this interesting species. From the point of view of the earth, either outcome is equally natural.

CHAPTER VIII

Ecologism into the Eighties: Collapse and Continuity

Was ecologism merely a fad of the sixties and seventies, a passing social excitement with no long-term effect on American attitudes? After all, economic growth is still an unchallengeable national policy; the public seems as fascinated with technology as ever (witness the computer boom); and the conservative administration of decidedly anti-ecologistic principles that was re-elected in the landslide victory of 1984 managed to keep its popularity virtually throughout its second term. Is the era of ecological thinking a mere historical parenthesis, now closed?

There can be little doubt that some retrenchment of public willingness to see things in terms of limits to growth has occurred, and along with it some backing off from some environmental goals. Indeed, it would be surprising if this were not the case. No radical social movement can escape the inevitable period of reaction, the reverse pendulum-swing that gives society a pause in which to consolidate, consider, and regroup.

The early successes of environmentalism were achieved nearly by acclamation. By the middle seventies, President Jimmy Carter had made "limits to growth" a leading theme of his administration. It clearly deeply influenced the way this thoughtful and well-read president viewed the world. But within a few years, when combined with other setbacks in domestic and foreign affairs, Carter's energy policy of limits and restraints led to a widespread public frustration with what was perceived as the administration's general inability to get things done. People were fed up with "limits"; they wanted "can-do."

The rejection of President Carter and this policy illustrates two important points about worldviews. One is that a worldview is a perceptual paradigm that substantially shapes the reality around it: it defines what facts are important, and tends to ignore or gloss over data which are anomalous. The second is that the worldview must, nevertheless, fit the real world reasonably well: it must work. If it does not, it will be subject to rejection or drastic alteration. In the case of the Carter administration, an unrealistic application of the limited-earth perspective to a specific problem—energy supplies—led directly to public reaction and reversion to the more familiar "endless growth" principles offered by his opponent in the 1980 election contest, Ronald Reagan.

Yet, in the context of retrenchment that followed that election, the ecological worldview did not disappear. The surprising fact is that despite two terms of energetic opposition by a well-liked president, protection and improvement of the environment has regathered public support for itself impressively. As we shall see, neither the oil "scarcity" fiasco of the Carter administration, nor the ideological opposition of the Reagan administration have been able to shake the basic American commitment to ecological health.

The Carter Collapse

The "limits to growth" theme appeared in the Carter administration most clearly in its energy program of 1977. This program included the famous "moral equivalent of war" speech, and emphasized conservation as the best answer to the petroleum shortages that were then developing. Two years later, the situation had only worsened. In his April 1979 speech to the nation, President Carter reiterated the unpleasant reality of the oil situation: "This is a painful step, and I'll give it to you straight. Each one of us will have to use less oil and pay more for it." The administration repeatedly emphasized that declines in amounts of oil were a simple fact of life: the world's oil was nonrenewable, we had already pumped too much of it, and henceforth we must get used to less of it. The President added the promise of a glamorous technological solution to the problem—the "synfuels" program. But payoff from this expensive program would (supposedly) come later; in the meantime, the President talked of sacrifice and austerity.

These were the times of jolting scarcities in the energy marketplace, and frightening increases in price. The winter of 1977 saw real discomfort in cold parts of the U.S. due to natural gas shortages. Long gasoline lines formed at service stations in the summer of 1979. In that year alone, prices for gasoline rose 35 percent, home heating oil 36 percent, and food 11 percent. By the end of that year, the per-barrel price for foreign oil had climbed an astonishing 850 percent from pre-embargo prices, to $24—and did not stop there.

These experiences might have been taken as strong verification of the idea of limits. Certainly the Carter administration interpreted them this way, employing, as Daniel Yergin noted, the "dramatic imagery" of physical shortages in order to sell its energy policy to the public.[1] But there were disquieting inconsistencies that undercut the public's willingness to suffer. It could not be forgotten that the original shortage, the Arab oil embargo of 1973-74, had been political, not geophysical, in origin. And when the oil companies began to record staggeringly huge profits in the later seventies, the logic of "public sacrifice" wore very thin. President Carter sought to convince his public that "the energy crisis is real," but people increasingly suspected corporate collusion and price-fixing as the actual causes, as a *Time* magazine poll reported in April of 1979.[2]

An article in the *Reader's Digest,* appearing in 1980, graphically stated these suspicions. It detailed how during the natural gas shortage of 1977, gas producers had pointed to declining reserves to explain the problem, and advocated deregulation of gas prices as the needed incentive for new exploration. But barely a few months after deregulation in 1978—far too soon for any substantial exploration or discovery—the American Gas Association president "proclaimed that the long-term shortage was over." The conclusion drawn by this article, and no doubt many of its millions of readers, was that the whole thing was rigged.[3]

Was there a physical shortage of oil and other energy sources? No single question could more dramatically pit the "limited earth" worldview against its traditionalist foe, the previously unquestioned American commitment to endless growth and development. Both sides seemed content with such a clear dichotomy—a rarity in centrist American politics. For many of the liberal camp, energy shortage and the correct response to it became an ideological shibboleth. And conservatives, with equal vigor, rejected the administration's ecologistic version of what was happening and what it meant.

In spite of the problems at the gas pumps, there was little evidence that oil was literally running out. It was certainly true that domestic *production* was slowing: total energy production in 1978, for instance, was about 3 percent less than in 1972.[4] And it was equally true that international politics had first shut off some supplies, then jacked their price up. But to describe these sociological and political problems as "too little oil left in the ground" was to ignore plain and readily available facts. However true the "limits of earth" concept might be as a general principle, it could not substitute for factual accuracy in dealing with a specific, short-term problem. The physical shortage of oil was indeed dramatic imagery—not literal truth. This was almost certainly the most important cause of the massive public rejection of the Carter approach to energy, and equally certainly a primary contributor to the following recision from the limited-earth worldview.

A book published by the Conservation Commission of the World Energy Conference in 1978, for instance, recorded the prevailing informed opinion (among oil companies, government agencies, and consulting firms) that the "estimated, recoverable, conventional oil reserves" of the world were probably around 240 to 260 gigatonnes. This very conservative figure is the equivalent of about 44 years of energy, at 1972 world energy-use levels *of all energy sources.*[5] (In the real world, where many sources of energy besides oil are used, this figure would represent supplies for far more years.) Other analysts placed the proven reserves of oil, natural gas, coal, and related resources at about a hundred years' worth at a steady growth rate, not counting potential reserves and as-yet undiscovered sources.[6] These are not huge margins, perhaps, but they hardly constitute "none left in the ground," and emphatically did not cause the shortages of the 1970s.

In addition, the paradoxical tendency of proven oil reserves has been to grow, not shrink. That is, as demand increases, exploration increases also, and new fields are discovered which add to the known quantity of available oil. Although in absolute terms there is always less oil with each passing year, in terms of known reserves there is usually more. "World production grew by 70 percent between 1966 and 1974; in the same period, world reserves grew by 80 percent."[7] Until the process of exploration stops yielding this kind of increase, it can hardly be concluded that oil supplies are running out.

○ ○ ○

Why did the Carter administration, and a healthy portion of the American public, ignore the contrary evidence? Was it willful, perhaps treasonous, misrepresentation? Was it a gigantic hoax foisted upon a gullible public by profit-gouging multinational oil companies?

I would suggest that it was a case of paradigm-induced blindness: a case where the general truth of the limited earth was of such prime importance to its adherents that specific policies had to be constructed in accord with it. The Carter administration was selling a worldview in which it (and many other Americans) passionately believed: the notion that American use of world resources was excessive, and that the only responsible and moral course for the world's greatest resource-consumer was to begin to cut back its appetite. This is the natural point of view for the ecologistically minded. It is a moral point of view, founded on a persuasive and profound vision of what the earth is and how it works. To many educated and well-intentioned people, the ecological goals of conservation, efficient energy use, and resource protection formed a single package with questions of pollution and population; all were matters requiring immediate national soul searching and attitude reform. The moralistic tone of the Carter energy program of 1977 underlines this point. However true the general principle of the limited earth may be, however, it led policy makers to ignore the near-term facts and to misrepresent the actual situation. Eventually, and inevitably, the public listened to other voices.

The most widely read dissenting voice was that of the Hudson Institute, which published its optimistic interpretation of the world situation in 1976 under the title *The Next 200 Years: A Scenario for America and the World.* Against the "current malaise" of American opinion—the prevailing limited-earth view—Herman Kahn and his fellow writers offered an upbeat, pro-growth, business-as-usual picture:

> The biggest difference between us is in our conclusion that it is both safer and more rewarding to move forward with caution and prudence on the present course than to try to stop or even to slow down generally.[8]

Kahn's group regarded resources as plentiful. And this book expressed confidence that the ecological dangers of continued industrialization would either be averted by new technology, or be acceptable trade-offs for increased prosperity.

> Most predictions of damage hundreds of years from now tend to be incorrect because they ignore the curative possibilities inherent in technological and economic progress. . . . Future progress might very well remove most of the dangers or costs.[9]

Nowhere does the great difference in attitude between pro-growth thinking and ecologistic thinking show more clearly than on this point. The Hudson Institute's faith in technology is immense. America has traditionally venerated its inventor-saints—the Edisons, Whitneys, Fultons, Bells. They are the semiholy icons of the national faith that science, technology, and Yankee ingenuity will build a bright and marvellous future, overcoming present problems by yet unimagined means. On issue after issue, *The Next 200 Years* refers to probable or possible technological innovations that will solve the problem, eliminate the pollution, create the new food, new space, new energy needed to avoid any change in civilization's current course. These even include such exotic options as floating ocean cities and orbital or interplanetary colonies (though these are not made the formal basis of prediction). Nowhere does this book betray a hint of the heavy sense of responsible creatureliness which governs ecologism; nowhere the heightened respect for the uncomprehended complexities of the ecosphere; nowhere the sense that unrestrained technology and industrialism can be a vicious cycle, spawning multiple problems with each singular "solution." Instead:

> We take the position that nearly every measurable environmental blight or hazard can be corrected by a combination of technology, a reasonable amount of money, sufficient time to make the required changes and (occasionally or temporarily) some (otherwise undesirable) self-restraint.[10]

The grudging twice-qualified nod to self-restraint speaks volumes. Restraint goes against the grain for traditional growth-oriented thinking. But it is the very essence of ecologism.

The administration of Ronald Reagan which was elected in 1980 embodied exactly these traditional values, and offered them in direct opposition to the Carter years of restraint and limitation. Continued growth would liberate new marvels of progress. Technology would solve our problems, if we let it. No radical change in our basic thinking was needed. The "Star Wars" defense proposal of 1983 (that a foolproof nuclear defense against the Soviets might be created out of ultra-new technologies placed in earth orbit) might be seen as another example of the tendency to place faith in technological solutions to human problems—particularly when the human problems are as frustrating and intractable as the nuclear dilemma has been. But even more fundamentally, Ronald Reagan was elected to prove, in the face of shortages, embargoes, and hostage crises, that Americans were not feeble, not incapable, and most of all not limited.

Candidate Reagan offered a different interpretation of the energy crisis. It was caused neither by physical limits of the earth, nor by the cupidity of oil companies. "Government regulations and production barriers are the two chief causes of the energy crisis we are now in," he proclaimed. The solution? In the words of a chief adviser, "Production, production, production!"[11] A Texas supporter phrased the contrast in fundamental beliefs succinctly: "This country did not conserve its way to greatness. It produced its way to greatness!"[12] In November of 1980, at least, a majority of Americans agreed.

○ ○ ○

That ecologism has passed an early peak is clear. That two terms of conservative and traditional national policies have been immensely popular is equally clear. But there is evidence that the public attitude lies, in reality, somewhere between the two relatively extreme positions suggested by the energy debate of the late seventies. Substantial portions of the ecological revolution have, far from being abandoned, actually been incorporated into the structure of unquestioned American beliefs. As such, they may awaken less fervor, and attract less notice. But they can be violated only at great political cost.

The most obvious evidence of ongoing environmentalism in American politics is seen in the fate of Interior Secretary James Watt, Environmental Protection Agency administrator Anne Gorsuch Burford, toxic-waste fund chief Rita Lavelle, and a score or two of lesser officials, all fired or retired under pressure for transgressing this new but important tenet of American belief. Watt was brought into the administration as an advocate of drilling and mining interests (he had been a lawyer for these interests, fighting environmental restrictions, as a member of the Denver-based Mountain States Legal Foundation). He was abrasive, even mocking, toward conservationist and preservationist sensibilities. His job was to divest the federal government of as much land as possible, open up as many resources to developers as possible, and (judging by the record of the EPA) to hobble, roll back, or undercut as many portions of the environmental protection code as possible. The rationale for all this was *production*. America was in a recession. She needed her oil, coal, gas, timber, and other reserves. And her industries needed to get on with their productive mission without the government "on their backs."

The result was massive public rejection. While the idea of unrestrained American productivity played well as a campaign slogan, in the real world it ran up against an almost equally strongly held value: basic protection of the environment. It seems that the Reagan administration simply misjudged the extent of its 1980 mandate. It drew the conclusion that Americans cared virtually not at all for the values of a clean and safe environment. But when the public saw that clean air and water standards were being ignored or reduced, that huge amounts of public lands were being given over to development, and that even toxic-waste cleanup was being mishandled in favor of polluters, it reacted strongly. A substantial degree of ecological consciousness had become a permanent part of the American value system.

It is true that a contrary conclusion might be reached upon the evidence of Watt's successor and former undersecretary, Donald Hodel. He has differed from Watt mainly in being more circumspect as he goes about the task of weakening environmental protection and representing industrial and commercial interests. His actions are continuations of Watt's growth-at-any-cost policies. He has supported oil drilling off the Pacific shore; recommended that part of the Arctic National Wildlife Refuge be opened for oil exploitation; destroyed the effectiveness of the Interior office charged with enforcing strip-mine regulation and reclamation (by cutting its budget by one-third and its staff by half, and neglecting to collect $150 million in fines, for instance); allowed drilling and strip mining to the detriment of adjacent national parks; and a litany of other actions promoting economic development over environmental protection.[13] The last of Reagan's and Hodel's stewardship promises no better: according to Barry Flam, Chief Forester of the Wilderness Society, the 1989 budget "represents a continued, deliberate move away from the land ethic, away from land stewardship, and away from increasingly valuable public uses of public lands."[14]

Yet, while this (very incomplete) tally of Hodel's attack on the environment is surely sad news, it is beside the main point. Hodel has succeeded where Watt failed because he has been able to keep an extremely low profile and even disguise his anti-environmental programs, not because the public supports them. Hodel's nearly Machiavellian manipulation of public sentiment regarding the Hetch Hetchy valley illustrates the lengths to which he has been driven to divert public attention from more basic issues. Hetch Hetchy is perhaps the most symbolic locale within the domain of the Interior Department: here John Muir led a bitter nationwide battle—one of environmentalism's first—to save the valley's stupendous beauty from inundation by a San Francisco municipal water project. Muir lost the battle in 1913, and ten years later a 430-foot earthen dam plunged the valley under hundreds of feet of water. In 1987, Secretary Hodel proposed to tear down the dam and restore the valley, at great cost and with unknown chances for success. It is difficult not to see this proposal as a red herring. For while environmentalists argue in amazement over the merits of the idea, the oil leases, the timber sales, and the failure to enforce environmental regulation continue unchanged.

Why does Secretary Hodel find it necessary to resort to such political flummery? Because the American public fundamentally disagrees with Hodel, Watt, and even Ronald Reagan on this issue. Despite nearly two full terms of Ronald Reagan's immense popularity (and engrained opposition to environmentalism), the measurable public support for protection of the environment has actually increased, and constitutes a solid majority. A scientific public opinion poll conducted by Cambridge Reports, for instance, has tracked rising support for the statement, "We must be prepared to sacrifice economic growth in order to preserve and protect the environment." In 1986 this position received 58 percent support, against a mere 19 percent affirmation of the opposite statement ("We must be prepared to sacrifice environmental quality for economic growth"). The equivalent figures in 1981 were 41 percent and 26 percent, respectively.[15] Six years of Reaganism, far from derailing support for the environment, oddly stimulated it.

Sociologist Riley E. Dunlap has reviewed masses of opinion polling data for the 1970s and 1980s, and has concluded that the trend is demonstrably in the direction of increased environmentalism. The collected evidence from major scientific polls over more than a decade consistently shows "a significant upturn in public concern for environmental quality during the Reagan Presidency":

> In general, it appears fair to conclude that, after suffering deterioration in the late 1970s, environmental quality has again become a "consensual" issue—as it was in the early 1970s.[16]

Other measurements confirm this conclusion. Memberships in almost all environmental organizations, often flat during the later 1970s, have steadily and steeply climbed. The Sierra Club's membership has more than doubled since 1980; in the year 1987 it grew by 8 percent to 425,926.[17] Other large organizations like the National Audubon Society, National Wildlife Federation, and Wilderness Society have done comparably well; while many of the small organizations have done even better.[18] The surprising conclusion must be that environmentalism is stronger than ever, despite its late-seventies setbacks.

It should be emphasized that this kind of broad-based political environmentalism is not the same thing as ecologism. The wider public has certainly never subscribed to the deeper radicalism of genuine ecologism or deep ecology. But it is nevertheless true that some degree of ecological consciousness has become a permanent part of the American value system. This wide support, while not philosophically deep, has shown itself to be perhaps emotionally deep: tough, resilient, indeed almost ineradicable.

One might speculate about the relation between this widespread environmentalism, which coexists easily with many traditional values, and the thoroughgoing ecologistic core. It would seem that there is, in the general public, a great deal of good will to be tapped, even if it is not strictly "correct" in its intellectual understandings. And it would seem, further, that a public thus committed to the environment might respond surprisingly to intelligent, realistic leadership, should such be offered from inside the core of the ecologistic movement, its thinkers and writers and activists. Though staggeringly much still remains to be done even to slow the rate of environmental deterioration, there exists substantial political willingness for the committed to mobilize.

○ ○ ○

What kind of a future will meet us, mobilized or not, environmentally enlightened or not? It seems to be a matter of surprising agreement that the economic and environmental *status quo ante* will not be the appropriate response. The changes in perception and value that began in the 1960s and 1970s and continued in the 1980s can be seen as a realistic response to a changing world. As such, they are unlikely to simply melt away like yesterday's fashions.

Even highly conservative analysts like Herman Kahn and the Hudson Institute have described a future that looks surprisingly like the one foreseen by their ecologistic opponents. Population will stabilize. Economies will achieve "more or less a steady state." The values that will be appropriate in these future cultures will be, as this conservative book describes them, remarkably similar to the ones presently held by ecologism: a commitment to "means which are also ends," in contrast to current orientation toward progress; creation of lasting, stable economies; attention to inward development, rather than outward "success" or consumerism; cultivation of diversity; religious and aesthetic connection between humans and with nature. In sum, Kahn and the other writers of *The Next 200 Years* foresee a postindustrial society with many attitudes "also often found in a primary or pre-industrial society." Such predicted attitudes match the quasiprimitivism sometimes found among the ecologically minded—the eastern religions, the Amerindian affectations, the interest in prehistoric hunter-gatherer societies, the "return to nature." On a deeper level, these represent a search for ways of thinking about the world which go beyond the strict, cerebral rationalism of the Western scientific-historical worldview, and recapture some of the lost wholeness of primitive thinking.

The Kahn report makes exactly the same point made by *The Limits to Growth.* Both see ecologistic values as the ones which will replace the competition, consumerism, and headlong change of the present. The significant difference is in the time scale: Kahn projects these values into the future, and advocates business as usual during a leisurely transition. Ecologism believes the moment has already come, and sees its values as the only answer to deadly attitudes that pollute and kill ecosystems, that divide humans into warring camps and threaten them with holocaust, and that rob the human kind of its membership in the natural household of earthly life. But the direction of change is identical: as the Club of Rome put it, "the transition from growth to global equilibrium."[19]

In fact, the minority which holds more thoroughgoing ecologistic principles may be stronger, in the long run, than is immediately evident. For one thing, it seems probable that events will favor their views. Energy may not in the short run become truly scarce, but it will never again be cheap. This fact will tend to force a conserving outlook, one which seeks not to use up and throw away, but to maintain existing goods and energy sources. This attitude need not be felt as the sort of monastic self-denial which the Carter administration seemed to call for. The CONAES report of 1980 showed that economic growth *can* go along with growing conservation, predicting that the ratio of energy input to Gross National Product in the U.S. could be reduced by one-half to two-thirds of its 1973 value, with no significant impact on real personal income.[20] Countries such as West Germany maintain a comparable or superior level of income and amenities with a much smaller proportional energy input; in other words, they are more efficient. In fact, according to the congressional Office of Technology Assessment, this has been the trend in the U.S. economy ever since 1973.[21] Such a change is only a first step away from the consume-and-pollute mentality, of course—but first steps can mark significant changes in direction.

Thus it appears that those who hold ecologistic beliefs may play a larger role in shaping public policy and public attitudes than the conservatism of the Reagan years would imply. If nothing else they will continue to provide a vocal minority opinion, offering an alternative to a culture's habitual way of doing business. This viewpoint will be creating styles of responding to ecological crisis which, it seems probable, more and more will find sensible, as problems of population, pollution, and scarcity stubbornly refuse to "just go away." As we have seen, a subtle process of assimilation has already begun to carry ecologistic attitudes into the national mind.

But as the sudden collapse of President Carter's energy-crisis consensus demonstrates, responses to ecological crisis will have to be accurately matched to the real world. Reliance on an ideology, even the attractive values of ecologism, will not be able to replace practical applicability and factual accuracy. There are many in the ecological movement who are well aware of this, and it is from these people, deeply involved in appropriate technology and nuts-and-bolts problem solving, that alternatives for the future will continue to come.

CHAPTER IX

Ecologism into the Eighties: Orthodoxy and Schism

The 1980s have seen a strong political challenge to environmentalism answered by a resurgence of popular support. But this wide public stream acts and thinks differently from those who hold to a fully ecological worldview. How has ecologism itself been affected by this paradoxical climate of hostility and opportunity? What has happened to the energies of the movement, once so full of variety and wonder, the profound and the silly, and experimentation in thought and lifestyle?

Some obvious landmarks might be observed. *CoEvolution Quarterly* transformed itself into the *Whole Earth Review,* with an express interest in computer software. The science of ecology took a turn toward traditional reductive analysis, downplaying both its traditions of holism and its new cybernetic models. Ecofeminism came of age, becoming a well-defined movement not always exactly in concert with other forms of ecologism. An important new book, *Deep Ecology,* laid out philosophical groundwork for much—though not all—of the movement. David Brower left the Friends of the Earth, amid controversy as he had left the Sierra Club nearly two decades before, to found a new group called Earth Island Institute.

In fact, an accelerating process of radicalization, controversy, and division has become typical of ecologism in the eighties. Perhaps the best illustration is the tremendous disagreement swirling around the group Earth First! It was informally founded in 1980 by five disgruntled environmentalists. Frustrated by their failure to get truly significant environmental progress in federal action despite the official friendliness of the Carter administration, they decided something more direct and more radical was called for. Inspired by Edward Abbey's novel *The Monkey Wrench Gang,* they began a program of guerrilla tactics, vigils, and nonviolent resistance against clearcutters, bulldozers, and the like. Soon one of the founders, Howie Wolke, was arrested and sentenced for the simple yet effective act of pulling out several miles of surveyors' stakes along a projected logging road. Others began "spiking" trees in timber-sale areas—randomly driving long nails into the trees, which would disable any chainsaw or millsaw. They called it "ecotage" or "monkeywrenching."

Edward Abbey approved mightily, contributing a "Forward" to Earth First!'s *Ecodefense: A Field Guide to Monkeywrenching.* Since the wilderness is the "homeland" not only of humans but of many species, he said, people have "the right and the obligation" to fight its destruction.[1] But the mainline environmentalists were scandalized by such tactics. Jay Hair, president of the National Wildlife Federation, proclaimed, "They are outlaws; they are terrorists; they have no right being considered environmentalists."[2] Cecil Andrus and a long list of solemn sages have agreed: Abbey and Earth First! are beyond the pale.

Yet they avow nonviolence; they follow Thoreau and Gandhi as much as Abbey. (Disagreement comes from mill workers who fear flying shrapnel from shattered sawblades.) Their claim is that the threat to the earth is too great for teacup committee work and patience—a claim that could easily be supported by reference to the standard environmentalist writings of Brower, Commoner, Hardin, and others. Sierra Club fundraisers and Wildlife Federation lobbyists have been quoting them for years. As Wolke said, "If nothing else, we aim to light a fire under the traditional conservation groups."[3] The emergence of Earth First! illustrates how the radicalism of ecologistic principles can push actions beyond the comfortable mainstream of political give and take.

In fact, the debate over Earth First!'s tactics is accompanied by a deeper conflict over its fundamental values. For in its eagerness to save wilderness from the ravages of human overpopulation, Earth First! has drawn the conclusion that whatever reduces population ought to be welcomed. In the most glaring example, an article entitled "Is AIDS the Answer to an Environmentalist's Prayer?," the author speculates that Gaia might be using this plague as a cybernetic correction to the threat posed by humanity: "AIDS may be Earth's own response to human-created environmental problems."[4] Earth First! takes the ecologistic principle of biocentrism most seriously. The well-being of threatened global environments and nonhuman species outweighs the importance of human well-being. Human die-off corrects the problem.

Not surprisingly, this attitude looks like madness and misanthropy to many other parts of the environmental movement. Earth First! has been branded "eco-fascist" for its flirtation with violence and its advocacy of death. Environmentalist Murray Bookchin, whose books have for decades encouraged ecological awareness, has recently conducted a rollicking epistolary battle with Earth First! and Edward Abbey, insisting that they have oversimplified the problem into a kind of "deep zoology. . . . reasoned out by analogy with fruit flies." For Bookchin, the problem is not merely one of sheer population numbers, but of the social and ethical systems employed by those numbers. "Social abuses . . . can only be corrected by social changes."[5] Even if AIDS did wipe out 25 percent of world population (as the Earth First! author hopefully forecasts) what good would it do if the remaining three-quarters continued to reproduce and pollute as before? Biocentrism does not necessarily require one to take sides against humanity or even civilization. It may instead require that civilization be adapted to fit into the biosphere—to recognize humankind's proper and even unique role within the community of species.

These controversies illustrate the rising factionalism of the ecological movement in the 1980s. There is a distinct change of mood. Rancor, name-calling, and schism have ceased to be uncommon. The ecologically inclined are laying down definitions. They are becoming more selective, critical, and consistent in their thinking. Some are following radical thoughts to their conclusions in radical actions.

And as a result, the rambunctious forward-ho of the early days, in which the straight and the weird, the correct and the confused, the diehard and the dilettante, all walked together in defense of the earth, is seldom encountered. It has given way to a more institutionalized approach, where various groups of the like-minded vie with each other over tactics, theories, and doctrines.

The effect is a relative one, of course: at the individual level many people participate in more than one branch of the movement without any difficulty. "Green" people still value nonviolence, tolerance, and consensus, and are often struggling to enact these values in the context of political action, disagreement, and compromise. It is surely not an easy task! But the overall drift is observable nonetheless. As it matures, the ecological movement is paying more and more attention to theory and fundamentals, leading it often to break with those who are, from another perspective, fellow-travellers.

○ ○ ○

Bill Devall and George Sessions' 1985 book *Deep Ecology: Living as if Nature Mattered* represents the ecological movement in this new phase. It has been widely read and reviewed; and I think it will be, for many, the definitive ecologistic expression for some time to come.

Deep Ecology repeats a now-familiar call. It advocates "a shift in worldview based upon a metaphysics consistent with ecological interconnectedness."[6] It opposes the majority tradition—in science, as mechanistic and limited; and in culture, as afflicted with the diseases of consumerism, materialism, individualism, and spiritual shallowness. This book gathers up past and present ecologistic themes, personalities, and ways of looking at the world, in a rich and detailed mosaic. It is a comprehensive statement which any student of the movement could profit by, and which any citizen of the earth should welcome.

But *Deep Ecology* also narrows some of the wide stream of ecologistic ideas. Perhaps as a result of the peculiar pressures of the 1980s—or perhaps as a natural evolution of ideas and movements—Devall and Sessions shift significantly both in attitude and in fundamental approach. Their version of ecologism is differently grounded; and it is more selective about who may be counted among the truly "deep."

Most significant is Devall and Sessions' partial abandonment of science as a principal foundation of the worldview. Repelled by materialist and reductionist science, they seek a foundation that "goes beyond the so-called factual scientific level to the level of the self and Earth wisdom." Though ecology is useful, it is too bound up in the materialist tradition of science to be trusted very far. Ecological science has at times provided a wider and deeper view of nature; but in general "deep ecologists" "have understood the need to go beyond the narrow definition of scientific data and look to their own consciousness."[7]

This approach, based on intuition, personal experience, and philosophical religion, comes in part from philosopher Arne Naess, who is quoted at length and with reverence.

> Ecology as a science does not ask what kind of a society would be the best for maintaining a particular ecosystem—that is considered a question for value theory, for politics, for ethics. As long as ecologists keep narrowly to their science, they do not ask such questions. What we need today is a tremendous expansion of ecological thinking in what I call *ecosophy*. . . . a shift from science to wisdom.
>
> This view is intuitive, as are all important views, in the sense that it can't be proven.[8]

Naess enumerates ecological "norms" and a list of "basic principles," all of which express the values and choices which we have explored in these chapters. Some of them are phrased in the semiscientific language we are used to: symbiosis, diversity, stability, equilibrium. But these norms "cannot be validated, of course, by the methodology of modern science based on its usual mechanistic assumptions and its very narrow definition of data." He goes on to say:

> The main point is that deep ecology has a religious component, fundamental intuitions that everyone must cultivate if he or she is to have a life based on values and not function like a computer.[9]

The basis of deep ecology, as Devall and Sessions emphasize, lies in the realm of personal experience and even mysticism. The science of ecology plays but a small role in making this worldview.

No doubt Naess and the deep ecologists are absolutely right: philosophy and science, "ought" and "is," can have nothing to say to each other—*if the discourse is framed in terms of the Cartesian dualism.* Within the classical Western definitions, these two realms are distinct.

The paradox of *Deep Ecology* is that it does not accept this dualism, yet unintentionally perpetuates it in its rejection of science in favor of intuition. The choice is framed as an either-or, with no attempt at transcending the dualistic frame. At the same time, however, the deep ecologists seek the goal of "integrating mind-body-spirit," and "development of mature persons who understand the immutable connection between themselves and the land community or person/planet."[10] It is hard to see how this integration can be achieved without transcending the necessity to choose between intuition and science. The more profound exponents of ecologism—notably Gregory Bateson—have stressed that intuition and science do not have to be kept separate from each other. That there is a single natural process that produces life, mind, rational knowledge, and living experience. And that it is the goal of ecological thinking to understand that unity, that single process.

It is not at all clear that what Stephen Gould has called the "bad habits" of reductive/mechanistic science (recently revived even in ecology) nor yet the worse habits of American politics (perennially obvious) provide real justification for giving up this unity, for this retreat to the safety of the mystical side of the dualistic teeter-totter. Bateson was well aware of these familiar extremes: scientific inability to talk about mind and purpose; mystic inability to connect its intuition with public knowledge about the physical world. To bridge the division, he offered what he called "cybernetic ethics," a monistic approach unifying science and ethics, based upon

> an epistemology rooted in improved science and in the obvious. I believe that these arguments are important at the present time of widespread skepticism—even that they are today as important as

> the testimony of those whose religious faith is based on inner light and "cosmic" experience. Indeed, the steadfast faith of an Einstein or a Whitehead is worth a thousand sanctimonious utterances.[11]

Bateson chooses his examples carefully: the scientist of relativity and the philosopher of a rigorously naturalistic view of the world as defined through its relations. He means to emphasize the importance of working *through* the natural world to affirmation and spirituality, not *around* it.

So one is left with questions for the deep ecologists. Why bring ecology into it at all, even in name? Where is the distinction between the two traditions of ecology (which is given the barest mention)? Where is the means of integration for those who may not have had the requisite intuition?

One might well speculate about a motive of defensiveness behind *Deep Ecology*'s retreat from science. We have examined a number of setbacks for the holistic tradition within scientific ecology. These, as we saw, were seldom complete reversals, but rather represented partial reformulation and qualification. In many cases they were also part of a larger tug-of-war between the two very well-entrenched (and historically legitimate) sides, holist and reductionist. These developments call for caution from ecologistic thinkers—for flexibility, humor, humility before the facts. But they do not call for abandoning the field.

The cultural situation of the mid-eighties also invites speculation about defensiveness. Under pressure from a conservative political atmosphere, has the ecological movement begun to see itself as somewhat beseiged, perhaps even as a movement paradoxically forced to stop moving, to draw itself together and throw up earthworks (one might say) for defense? The previous chapter suggests that, in fact, this would be an ironically inaccurate reading of the situation: popular environmentalism has not diminished. Yet the question here is perception; and the noisy anti-environmentalism of James Watt and the early Reagan administration may well have awakened a defensive mentality. It may be, therefore, that *Deep Ecology*'s retreat from science is part of a retrenchment in search of an absolutely safe foundation.

To abandon the public form of knowledge (science) in favor of an unassailable philosophical/experiential basis may or may not be the way future ecologistic thought will develop. But it is strangely reminiscent of a similar moment in the mid-nineteenth century, when the rational religion of orthodox Christianity—the religion of Paley—found itself at odds with biology and other developing sciences. One result was a movement *within*, a retreat to the safe precincts of a religion based on feeling and subjectivity (most notably in the theology of Friedrich Schleiermacher). It is a strategic redefinition that seems to lose as much as it gains.

If this move signals the reseparation of the knowing intellect from the feeling and experiencing self, then it would seem to choke off the very lifeblood of the ecological worldview. I think that the deep ecologists would probably argue that they are not thus strangling the movement. But they have surely at least greatly redefined and attenuated the connection between its heart and its head.

○ ○ ○

Another, probably related, feature of the changes represented by *Deep Ecology* is visible in the authors' attitude toward other parts of the ecological movement. In its desire to lay a secure and rigorous philosophical foundation, *Deep Ecology* sometimes creates a kind of orthodoxy by which other parts of the movement are judged and found wanting. *Deep Ecology* rejects some of the people and ideas which, despite their occasional inconsistencies, have been in practice an integral part of ecologism.

It is not that Devall and Sessions are at all narrow minded or ungenerous; they offer a prodigious array of quotes and friendly mentions. But they also show a tendency to criticize other writers and thinkers who seem not to be "deep" enough—for failing, it almost seems, to adopt the full and complete doctrine. Wendell Berry provides a mild example. While he receives some sincere praise (an "eloquent spokesman," "one of the first," etc.), each encomium is followed by an exception:

> But . . . he fails to see the ecological necessity. . . .

> But Berry misses the deeper ecological reasons. . . .
>
> But. . . . Berry apparently fails to see the contradiction and falls short of deep ecological awareness.[12]

Is this merely intelligent, balanced criticism, or is it the application of an *a priori* set of tenets to a writer who has never heard of them? One senses a bit of both in these passages.

The authors of *Deep Ecology* take unmistakable aim, moreover, at one important group within the ecologistic movement: those who pursue a cybernetic interpretation of nature. They object to it on virtually every level; and in doing so (I would suggest) they reject and misunderstand much that is central to ecologism. In effect, they create a schism. First, *Deep Ecology* rejects cybernetic science:

> Much of scientific ecological theory is based on cybernetics systems theory—a continuation of the Cartesian seventeenth-century view of the universe as a machine—and should be held suspect for that reason.[13]

If it were true, this would indeed be a damning criticism. But, as I hope the preceding chapters have shown, the very point of cybernetic ecology is that it *undercuts* the Cartesian conception of reality. Simply because cybernetics derives from engineering and is associated with computers does not mean that it is mechanistic. Cybernetics redefines the world in terms of process, information, and relation —in total contrast to the Newtonian world of atomistic "matter in motion." By seeing matter and mind, life and nonlife, as different aspects of a single process, cybernetic philosophy unifies the dual sides of the Cartesian conception. The cybernetic universe is alive, generative, and creative—while the mechanical universe of Newton and Descartes is dead.[14]

Secondly, based on this misunderstanding of cybernetics, Devall and Sessions associate it with a so-called "New Age/Aquarian Conspiracy," described as an extreme program of human-centered growth and domineering control over nature. This unwholesome group evidently believes that "Engineering, coupled with cybernetics systems and information theory, can provide a purely technological solution to the world's ills." And furthermore, according to Devall and Sessions:

> The major metaphor of New Age is that the Earth is a spaceship and technologically advanced humans have a destiny to become "co-pilots" of the spaceship.[15]

This interpretation of "spaceship earth" is especially bewildering. Evidently the technological echo of the phrase horrifies them. But as is evident in the quotation we looked at earlier (in Chapter 1) from Commander Borman, the point was not control over earth, but affection for it; not that earth belonged to the viewer, but that the faraway human understood for the first time how *he* belonged to the earth. No doubt some few (such as those quoted by Devall and Sessions) do find the other message in the phrase and its familiar picture—but not many. Rather, the "spaceship earth" metaphor usually appears as part of a strong ecological message of conserving and caring for the planet. One example, taken almost at random, is found in E.F. Schumacher's powerfully ecologistic *Small Is Beautiful.* Schumacher makes the point that natural resources are, economically speaking, more like "capital" than "income." Then he makes the application:

> A businessman would not consider a firm to have solved its problems . . . if he saw that it was rapidly consuming its capital. How, then, could we overlook this vital fact when it comes to that very big firm, the economy of Spaceship Earth[?][16]

Schumacher reaches for the spaceship expression almost effortlessly—its ecologistic meaning is so familiar that he uses it virtually as a cliché, and despite its slight incongruity with the principal economic metaphor of the passage.

So it is odd indeed that *Deep Ecology* repeatedly associates the "spaceship earth" metaphor not only with New Age technocentrism, but also with the desire to abandon earth via spaceflight, with the implication that we therefore need not worry about polluting and exhausting the planet. Rare and quirky indeed is such a use of this image and its associated picture from space: the blue-green capsule of life that is such a poignant reminder to cherish above all else this rare, self-enclosed, hospitable place.

○ ○ ○

As a result of the above interpretations, *Deep Ecology* excludes a large part of the ecological movement from full membership. For instance, part of Devall and Sessions' criticism of cybernetic ecology gets directed at that archengineer, Buckminster Fuller, whom they all but denounce. Perhaps some of this critique is well founded; Fuller did bring out the machine side of the spaceship metaphor. Yet he also promoted whole-earth thinking, an essential ecological attitude that tried to see the connections between earth's many human-caused problems. Surely this side of Fuller need not be cast completely out of the ecological church.

Similarly, the blithe spirits at *Whole Earth Catalog* and *CoEvolution Quarterly* (now *Whole Earth Review*), are distinctly "out" to deep ecology, evidently part of this technology-crazed conspiracy. For they are tool-oriented, interested in spaceflight and computers. And as editor Stewart Brand remarked in the Fall 1983 issue: "The *Whole Earth Review* and *CoEvolution Quarterly* are godchildren . . . of Buckminster Fuller." Yet in this very issue appears Gary Snyder's powerful essay "Good, Wild, Sacred." And if one glances over the concerns of *CoEvolution Quarterly* during its ten-year history (1974 to 1984), one finds sustained attention given to virtually all aspects of fundamental ecologism or deep ecology: voluntary simplicity. Preserving agricultural and regional diversity. Peacemaking. Gaia. Bioregions. "How to liberate mind & body and protect endangered species (including ourselves) from pathogenic industrial civilization."[17] And so on: by Devall and Sessions' own standards and declarations, this is unquestionably deep ecology. What's a philosopher to do with this messy mingling, this unholy yoking-together of the deep and the denounced?

For in fact the ecological movement has historically included many in it who might not be pure (or "deep") enough to satisfy Devall and Sessions' rigor. *Deep Ecology* seems to pry apart what has been a diverse, rambunctious, squabbling unity. This unity has had to do with openness to change, excitement about the future; eagerness to celebrate the mind and the body, to embrace intellect and biology, to co-adapt nature and technology. The emphasis has always been on the "and."

It is this messy "and" quality that is sometimes missing from *Deep Ecology*, which often prefers the excluding "but." I suspect this fact represents an often-observed phenomenon, that it is the fate of movements to harden eventually into institutions. When they do, they begin to define their shibboleths, to split into factions, to become more consistent but less vigorous.

The ecologistic worldview can perhaps be visualized as an ecosystem of ideas: an assemblage of images, scientific facts and principles, metaphors, feelings, father- and mother-figures, which have entered into a synergetic pattern of relations, somehow all playing off and with and against each other, and making a whole of it. (The reappearance of this whole, this cluster of ideas, in unexpected places—such as Walter Cannon's writings—strongly suggests some kind of affinity, some tendency to converge.) Such a coming together brings in many diverse and contradictory elements. These, in the movement phase, are not cast out. Rather, they are accepted for whatever contribution they make. Inconsistencies are not much worried about. Tensions abound. The real world of ideas in use is approximate, everchanging, fertile, messy. It tends to be suggestive and inclusive, rather than exhaustive and exclusive.

As we have already seen, the self-contradictions of the ecologistic worldview are many. Is the earth frail and endangered, or a robust, unshakable under-reality? Is change or stability the basic condition of life? How shall diverse individuals be fully respected and actualized, while at the same time obeying the imperatives of the larger whole? Is ecologism anarchistic or authoritarian, individualistic or collectivist? And most paradoxically, is human nature, nature? Or is it artifice? Or even, anti-nature? The ecologistic answers to these questions fall on both sides of each fence. The problems are not resolved in any systematic way. But they are less pressing in practice than in theory, as I suggested in the chapter on ethics. I think that visions, metaphors, analogies, and yes, even intuitions form a common atmosphere for those within the worldview, and that such people collectively breathe in a subtle sense for how to proceed when faced with concrete decisions.

In an unconscious homage to this sloppiness of ideas, *Deep Ecology* uses the Gaia metaphor surprisingly freely. The Gaia Hypothesis is, of course, strongly implicated in the cybernetic view—its very basis is a concept of global information-processing. As I have said, Gaia exemplifies not only the ecological superorganism, but the ecological supercybernetic mind as well. Strangely then, *Deep Ecology* uses the Gaia metaphor as freely as does the rest of the ecological movement.

However, Devall and Sessions wish to split the image in two: they criticize "the idea of Gaia treated as a scientific theory rather than myth."[18] They aim to keep the myth, but throw out the science by which Lovelock generated it. This dividing-in-two could serve as an emblem of *Deep Ecology*'s underlying problem, or at the least its distance from the rest of ecologism. For as I have emphasized, a basic (although unstated) premise of ecologism has been to unify the mythic and the scientific, the felt and the factual. Only if there is some kind of literal truth to Gaia can the image fully touch us and move us. We can accommodate scientific hedging, questioning, but-if-ing—as long as there remains some measure of sober, respectable, usable truth in the possibility. Only then do we feel connected into the *whole* world: the world of fact *and* vision, intellect *and* experience.

❍ ❍ ❍

One more example will establish the point that a certain doctrinaire quality, a spirit of exclusion rather than inclusion, has begun to make itself felt among the most serious and committed proponents of the ecological worldview.

In an essay recently appearing in the journal *Environmental Ethics,* University of Wisconsin philosopher Jim Cheney examines "Eco-Feminism and Deep Ecology."[19] This is a scholarly article, and so might be expected to split some hairs and make some rigorous logical categories. These are the stuff of clear thinking, and the article certainly makes some useful analytical distinctions on several important issues. Yet it also is curiously and in some instances unnecessarily trapped in the same either-or mode that characterizes some features of *Deep Ecology*.

Cheney observes that while ecofeminism has much in common with the deep ecology movement, the latter falls short because its basis is incompletely feminized—or, in other words, is stuck in a masculine mode of thought. The particular point at which this "masculine voice" is most audible is in the holism advocated by Naess and the deep ecologists. Cheney offers a striking interpretation of holism as a sort of secret twin of its opposite, atomism. This "male-authored holism" offers an oceanic fusion of the self with the whole—what we have encountered often as the organismic analogy and the myth of the superorganism.

> There is a strong tendency on the part of male theorists to understand networks of defining relationships on the model of an expansion of the self to the boundaries of the whole. This is, to be sure, a way of overcoming alienation, and as a way of having one's cake and eating it too, it can't be beat: one overcomes alienation from the other by absorbing the other into the self. There is, however, no respecting of the other *as* other.[20]

Such holism often gives a vaguely feminist aura or feeling to deep ecology, says Cheney; yet, citing other feminists like Claudia Card and Carol Gilligan, he labels the resemblance superficial and, in fact, wrong. This kind of oceanic holism is the desperate reaction of a typically masculine mind: "Fusion is a move toward health from a norm (an atomism) which is itself pathological—a move toward health, which, however, carries the sickness of atomism with it." It is but "the self writ large in the biosphere," an "atomism of one."[21]

The way to escape this dilemma is through the genuinely transforming feminist version of holism: a vision of the community as a network of particular connections, "women's web-like relations."[22] Cheney illustrates the difference between the correct and incorrect ideas through the "gift economy," a community of mutual responsibility in which care is not a market commodity, but a free act which binds the self to the whole and defines it as part of the community. This is not a metaphysical definition but an operational one emerging out of the daily routine of caring for others and receiving care back from them. It occupies a middle ground between the somewhat abstract extremes of the masculine duality ("either atomistically defined selves who are strangers to one another or one gigantic self").[23]

Probably the most useful feature of these distinctions lies in their ability to resolve the ecologistic value conflict between the individual (the values of diversity, self-realization, self-reliance) and the whole (the overarching values of unity and connectedness, the stability of the group or ecosystem). As we have seen, ecologism offers little specific guidance in resolving this conflict. Ecofeminism's middle ground, however, deals with the problem nicely, seeing the whole as an actual web-like community of mutually responsible individuals. As a real community, not an abstract conception, it is able to let answers to such dilemmas emerge from the group in the form of consensus, through time, patience, and caring mutual adjustment. "Consensual decision making takes relations seriously; it is a method of inclusion, and it is concerned to preserve community."[24] This insight is a real breakthrough, offering both theoretical and practical responses to a difficult ecologistic problem. And it accurately reflects the practice of many ecologistic, Green, and left-wing groups, as well as religious groups like the Quakers. All have found consensus decision making to be a genuine expression of a whole community that leaves no member feeling excluded, ignored, or voted down.

Thus the ecofeminist critique of deep ecology makes valuable contributions to the ecological worldview. Yet even Cheney, oddly enough, falls into the trap of what might be called intellectual imperialism—the insistence that there is one correct position, and "I've got it." For he offers the web-like community in *opposition* to the other holistic visions, not as a complement to them. He insists that the oceanic fusion of the organismic analogy is a sickness, leading to "ecological totalitarianism," just as deep ecology itself is doomed to be (quoting Ariel Salleh) "'simply another self-congratulatory reformist move' rather than a genuine feminist 'transvaluation of values.'"[25]

As this book has made clear, the myth of the superorganism has historically been an essential one for the ecological worldview. To cast it out in such uncompromising terms seems oddly out of keeping with a consensual, feminist, inclusive, community-oriented way of thinking. The community in question is the movement of ecologism, with its writers, both living and dead, and all its present-day followers, leaders, and hangers-on. Under the wide canopy of ecologism there are many myths, analogies, nature parables, principles, and values. They are not in strict logical conformity with each other, though they do achieve an interesting sort of synergism. As I have pointed out, many of the principal myths express the unity of life in a variety of ways, each presenting the mystery of the one and the many in a different light. While the superorganism does indeed fuse individuals into one, it may do so in different ways: at one extreme, it sees the parts joined in homeostasis as in a single body; yet at another, the planetary being Gaia shows a special ability to incorporate very real, very independent parts. "Fusion" may be a misleading term in this instance. And other important myths explore other kinds of unity: the symbiotic pair emphasizes partnership, and the climax ecosystem incorporates at least four kinds of unity, depending on the interpretation (it might be a single organism, a system of bio-energetics, an information-processing unit, or a web-like community).

By pointing out an arena in which this last interpretation—the web—is particularly well understood through ecofeminism, Cheney makes an important contribution. But by insisting on an either-or application of this idea, he largely misses the opportunity to locate it within, instead of against, the rest of the spectrum of ecologistic ideas.[26] The same tendency toward exclusivity and schism that affected *Deep Ecology* reappears here.

These are metaphors, after all, not statements of fact. Each says what life is *like,* and only indirectly what life *is.* Because of this, there is no reason not to hold various pictures simultaneously—each of them potentially useful, to be chosen as the occasion warrants. Sometimes, an individual *is* an individual. Other times, the organic wholeness of the group or the natural system will present itself as the relevant model. And other times yet, the web-like community of relations.

The necessary distinctions of clear thinking need not accompany a controversializing or combative attitude. There is room in ecologism for a richness of metaphors, an abundance of pictures, a forest full of creatures in varying and complex interrelation.

○ ○ ○

It may be that ecologism has, indeed, passed out of its movement phase, and that the chilling atmosphere of the eighties will permanently freeze the various parts of the movement into separate, vaguely alienated groups. If so, the ecofeminists and the deep ecologists will certainly be influential branches, continuing to offer to a needy world much of the message of ecologism. Or new events, new developments in information-, cybernetics-, and systems-science, might thaw out the sloppy coalition, and carry the movement forward in disorganized delight yet further. These outcomes are certainly beyond my ability to predict.

But in either case, the ecological worldview has found its way into the eighties with surprising vigor. It seems certain that the physical realities of the earth, as well as the psychological and spiritual needs of its human inhabitants, will guarantee an interesting and creative future for the worldview of ecology.

CHAPTER X

The Last Dethronement

To admit that humankind's prized possession, the mind, is not uniquely human is to abandon one of the last strongholds of the Western human-centered view of the cosmos. A moment's reflection will recall the outer circles of defense which have long since fallen. The Greco-Christian tradition created a universe whose furthest star and nearest object were all arranged symmetrically around the human being, enthroned at the feet of the invisible Father. But step by step this fantasy of a unique and privileged position in the cosmos has been destroyed. Copernicus demoted the earth from standing at the motionless center of the heavens to merely travelling along one undistinguished orbital path among many. Other astronomers followed by removing the solar system itself to an obscure place toward the edge of an average galaxy. What was left of the exalted human status was undermined further by Darwin, who ruined the pleasant fiction that *Homo sapiens* as a biological species was separate unto itself, disdainfully unrelated to the "lesser" brutes sharing the earth. Ecology has demonstrated that even present-day individuals, and even highly advanced civilizations, are not separable from the natural world.

Cybernetic ecology completes the dethronement of the human ego. The last step in this regress is the admission that even humanity's own mind is no conferrer of unique status. The rational consciousness is only one of a long list of other consciousnesses and mentalities, some of them in body tissues and genes, some in other organisms, some in the environment. Human thought, feeling, and mental achievement are products of a natural evolution and development as surely as are birds' nests or bees' wings. The human mind is only a more focussed and intense form of the universal process of nature. There is no place left to hide: the conscious animal must accept that it is at home on earth, and that earth, in some profound way, is at home in it.

Ecologism begins with practical, even mundane, problems of pollution, population, and industrialization. But it ends with unexpectedly exalted and far-reaching reconsiderations of the human place in the scheme of things. President Nixon, it will be remembered, called for a "basic reform" in perceptions, knowledge, and attitudes. Those who have followed up his call have discovered just how far the reform must finally go. Alan Watts, for example, has recognized that ecology requires "a fundamental revision of the very roots of our common sense." Such a revision could hardly stop with matters of technique and a few good pollution laws, as the President may have thought. It must inevitably reach down into the very heart of our sense of self and reality: "It is primarily a way of life in which the original sense of the seamless unity of nature is restored without the loss of individual consciousness."[1] Watts and other ecologistic writers come finally to focus on the very identity of the individual human being, challenging him or her to rediscover a connectedness beyond the constricted limits of personality and body.

Paul Shepard points out the way our language and culture have obscured this seamless unity by training us to see ourselves as individual "things" isolated from the rest of the world. But our new understanding of nature calls for something else—"a kind of vision across boundaries":

> The epidermis of the skin is ecologically like a pond surface or a forest soil, not a shell so much as a delicate interpenetration. It reveals the self ennobled and extended rather than threatened[,] as part of the landscape and the ecosystem, because the beauty and complexity of nature are continuous with ourselves. . . .
>
> [To see ourselves this way] means nothing less than a shift in our whole frame of reference and our attitude towards life itself, a wider

> perception of the landscape as a creative, harmonious being where relationships of things are as real as the things. Without losing our sense of a great human destiny and without intellectual surrender, we must affirm that the world is a being, a part of our own body.[2]

○ ○ ○

Watts has been a spokesperson for this point of view in the West for several decades now. For him, the rise of ecologism has been merely an unexpected avenue into the familiar home truths of Eastern religion, which he has for so long tried to interpret to a skeptical audience. This ecological avenue, it might be added, has surely been the most Western of routes: one taken with measurements at every step, and hard logic about results and tradeoffs, and scientific papers on every conceivable aspect. But its terminus is nonetheless an exotic one. In a long and unexpected cycle, centuries of scientific thought and millennia of religious and philosophical tradition have brought the West back to the place of beginning. And there they have found a smiling Buddha, asking "Why so anxious? Why so lonely?"

Watts discusses the basic unity of the human and the world as a figure/ground relation. If one draws a circle, it defines not one but *two* edges: that of the enclosed figure, and that of the surrounding field. Yet there is only one boundary. The point is that, sharing a boundary, the figure and the field cannot move or exist, except mutually. When I move, the universe moves with me. My shape and existence imply the shape and existence of the surrounding world—and vice versa. Each is an aspect of the other.

If the polarity of human/world is only an aspect of a more basic unity, then the naturalness of our being and doing is guaranteed. What humans do—thinking, building, reproducing—is in every point an activity of the cosmos. And perhaps most curious of all to consider: in human consciousness, the universe is knowing itself.

> We have shattered the myth of the Fully Automatic Universe where human consciousness and intelligence are a fluke in the midst of boundless stupidity. For if the behavior of an organism is intelligible

> only in relation to its environment, intelligent behavior implies an intelligent environment. . . . Surely all forms of life, including man, must be understood as "symptoms" of the earth, the solar system, and the galaxy—in which case we cannot escape the conclusion that the galaxy is intelligent.[3]

Or to put it the other way around, our knowledge of the world is finally self-knowledge. This cycle, too, has begun to complete itself. The aggressively objective practice of science, pursued with such steely vigor, such determined exclusion of the observer, has penetrated the natural world and come out the other side, where it finds itself finding itself. In physics and biology alike it is forced to admit the unity of observer and observed, the primacy of holistic fields over reductive dissection, and the inseparability of the human from the natural. The analysis of nature builds to an unexpected climax, in which researchers discover themselves to be the ultimate subject of their own studies. They are the world's mind, thinking about itself.

Laying down, at last, the myth of human uniqueness and separateness, the Western mind opens to modes of thought and being which have been, till now, wholly alien. Like the religions of the East, which have long been strange and inscrutable to us precisely because of this, they require a letting-go of the ego, a willingness for the universe to flow through the individual, and for the individual to flow with and into the universe. The most extreme challenge to the "common sense" of the West surely lies in this insistence that, after all, the unit-person on which we base our entire scheme of personal, cultural, and global sin and salvation, may not be as detached and self-contained as we have imagined.

These are hard sayings, and hard doings, for proud, brittle-shelled "Western man" (necessarily, and symbolically, male). He fears that the least crack in his precious boundary of self will lead to a mindless zombieism, to a slack-jawed acquiescence to the whims of nature or the tyrannies of tradition. Yet these fears cannot protect him from the ecological knowledge that he is not, indeed, separate, and cannot be.

Wisdom and the Faustian Tradition

The fictional character of Faust has often been taken as the embodiment of this Western man. As portrayed by Goethe, Faust represents that striving for self-realization, that heaven-storming thirst for knowledge and its mate, power, which since the Renaissance have driven Europeans into world dominion. Faust is the epitome of the human as privileged being for whom all the world is an arena in which to act heroically.

Faust is no petty criminal, selling cheap his soul; he gains a long earthly career, filled with learning and achievement. His most typical act, undertaken as master engineer for a Dutch prince, is to regain lands from the sea—re-enacting, perhaps, the divine creation in Genesis. It is an act of statesmanship and humanity. Above all, it serves the cause of progress.

Life under a Western sense of history is goal directed and linear. The hero of the old worldview, whether warrior, explorer, scientist, or entrepreneur, was like the Christian pilgrim—travelling the road to the divine or secular City of God. The hero's heart was always Faustian: focussed on attainment and success, ruthless toward obstacles, relentlessly moving forward. Whether expressed in the New Jerusalem of the Bible, or the Alabaster City of American myth, or the Socialist Utopia of Marxist belief, the End is a realm where nature is triumphed over, or transcended altogether.

In contrast, the wider view implicit in ecology stresses the overwhelming importance of nurture, birth, and renewal, in the long term; in other words, the paramount and inescapable cycles of nature. In this it tends to countervail even that most essential Western and Faustian belief, *history.* The familiar Western view of time as an arrow, speeding toward the goal of Progress, does not express time as experienced in nature, where the long, slow seasons repeat themselves immemorially, and where the important events are the known and expected recursions of fertility, birth, and death.

Rather than, like Faust, triumphing over nature, the hero of the new worldview would embody it, live it out, express its fullness and completion. In fact, the hero of the new worldview is just as likely to be a heroine—a nurturing, rooted person, living and dying in the same rustic place, bringing life out of the earth, and returning it again. This surprising idea has been a subtheme in Western culture, a renegade view. It peeps out of Rousseau's writings, and motivates the imaginatively potent poetry, art, and music of the Romantics.

Now this subtheme comes to the fore to answer a need which has been building for centuries. For European Man (as "he" would have seen himself) has travelled the road from being the Creator's conquering heir, to being the orphan of a distant (or altogether absent) father-god. Now, as if in the plot of a Dickens novel, he is suddenly seen as the beloved child of a rediscovered mother, the Earth. The relief, the sense of recovered identity, the restored inheritance, are palpable, emotional, and immense. The science of ecology has triggered this rediscovery, though the scientist of course does not speak this mythological language. The scientist refers to the "web of life" or the "matrix" of interdependencies and forces within which humans live. But after all, "matrix" turns out to be just a fancy word for womb or mother. The ecological awakening, at least in part, expresses the deeply feminine side of human nature, that side which has been largely lost to our official cultural identity.

The genius of Western civilization has been belief in the power and importance of the human. But it has led at last to belief in power alone, worship of technique at the expense of all other considerations, and division of humanity into isolated cells of will and action. Respecting nothing but the imagined limitlessness of human power, and being aware of no existence but its own shrunken self, *Homo "sapiens"* has finally come to the brink of destroying the natural web of life which nourishes it. Its crazed ideas of self and world have come near to causing ruin.

The achievements and ambitions of our Faustian culture are not to be lightly dismissed: no serious consideration of our predicament can pretend to wish away the accomplishments of reason and science. Even the most ecological reformers accept medical care, warm house and clothes, the thrilling expansion of knowledge even to the outer fringes of the universe and the deepest recesses of the cell. Yet such knowledge and power alone have proven to be inadequate. Knowledge and power alone, in fact, have proven to be fatal. They need to be informed with that wider perspective which Watts and Shepard and others have called for: something often called, simply, wisdom.

Is it not inevitable that the eagerly sought overdevelopment of industry and armament will finally, eventually, bring the taxed, polluted, atomized citizen to say, "enough"? Action, growth, technology for their own sake will eventually force a reckoning. In the recent rejection of some forms of technology like the SST and nuclear power, this reckoning is starting to happen; it is the voice of experience. And once the prince, president, industrialist—or citizen—stops to count the cost, he or she is no longer Faust, but a person on the way to salvation. (Goethe clearly shows the original Faust making just this transition.)

It may be that Western civilization has experienced its Faustian phase as a kind of adolescence. This is mere metaphor, of course, but it suggests nicely the possible significance of the gradual spread of ecological "wisdom": the onset of maturity, the sober reconsideration of all this exuberance of things and frenzy of action. A mature philosophy of ecological wisdom must replace the crass, hypermasculinized values of Faust. Without these values of nurture and peace, our species and our biosphere do not seem to have much future.

○ ○ ○

Walter Cannon chose the word "wisdom" for his title, more than fifty years ago. He chose it carefully, suggesting that something more than mere medical knowledge was glimpsed in the homeostatic workings of a living being. The "Wisdom of the Body" turned out to be that same wisdom found in the ecosystem by the Leopolds and Carsons, the Ehrlichs and Commoners of the ecological movement: the wisdom to recognize the unity of humanity with itself and with nature; to respect the limits of organic, environmental, and social systems; to cooperate; and to aim for stability, not endless growth, according to the natural example. Cannon's chosen word may be taken as a one-word summary of the four concepts around which ecologistic thinking has grown.

Wisdom contrasts not only with folly, but also with "mere" knowledge. A factual grasp is not enough, if it only sees the surface features or immediate causes and effects. Wisdom sees further, and deeper; it works within a larger perspective, and knows that actions may lead to distant and unforeseen results. There is patience in wisdom, along with the suggestion of maturity, ripeness, lack of hurry, long-sightedness. Wisdom involves not just intellect but character. In the words of Lewis Perelman, "Wisdom is the function of a mind that is respectful of its own boundary and processes."[4] It is knowledge not cut off from the contexts that create knowledge.

In concert with many ecological thinkers, E.F. Schumacher pivots much of his argument around this contrast of knowledge and wisdom.

> The greatest danger invariably arises from the ruthless application, on a vast scale, of partial knowledge such as we are currently witnessing in the application of nuclear energy, of the new chemistry in agriculture, of transportation technology, and countless other things.
>
> Economically, our wrong living consists primarily in systematically cultivating greed and envy and thus building up a vast array of totally unwarrantable wants.
>
> From an economic point of view, the central concept of wisdom is permanence.
>
> Wisdom demands a new orientation of science and technology towards the organic, the gentle, the nonviolent, the elegant and beautiful.[5]

"The alternative to this ecological insanity," as Professor Perelman says, is *wisdom.* The household of nature, the *oikos,* is the dwelling-place of man and woman—from some perspectives it is more like their very body; it may even be their very mind, or they its mind. But it takes more than knowledge to see this; it takes patience, humility, and lovingkindness. It takes an ability to extend love of self to love of the whole.

The particular ideas that are presented in this book as the components of the ecological worldview are ideas that deny the division and partial-sightedness of our Faustian tradition. They propose instead to use science, reason, imagination, emotion, and a renewed sense of the sacred in an attempt to reconstruct the whole of which we have tended to lose sight. The wisdom of the whole is that nothing is superfluous, no life insignificant, no living act separate from the web of all lives, no end separate from its means. And the wisdom of the whole is, further, that without understanding the unity beneath the diversity, without understanding the poise and balance and completeness of nature, the irresponsible acts of the conscious animal could destroy all.

This wisdom has been movingly expressed by many ecological poets and prophets who have been glanced at and borrowed from and analyzed in these chapters, but by none perhaps more eloquently than the novelist Ursula Le Guin. If wisdom is the reward of patience and age, then this passage from *A Wizard of Earthsea* well embodies it. These are only the words of imaginary characters in a paperback fiction—an elderly magician, speaking to a young man eager for power. Yet their message speaks the heart of the ecological worldview.

> Presently the mage said, speaking softly, "Do you see, Arren, how an act is not, as young men think, like a rock that one picks up and throws, and it hits or misses, and that's the end of it. When that rock is lifted, the earth is lighter; the hand that bears it heavier. When it is thrown, the circuits of the stars respond, and where it strikes or falls the universe is changed. On every act the balance of the whole depends. The winds and seas, the powers of water and earth and light, all that these do, and all that the beasts and green things do, is well done, and rightly done. All these act within the Equilibrium. From the hurricane and the great whale's sounding to the fall of a dry leaf and the gnat's flight, all they do is done within the balance of the whole. But we, insofar as we have power over the world and over one another, we must *learn* to do what the leaf and the whale and the wind do of their own nature. We must learn to keep the balance. Having intelligence, we must not act in ignorance. Having choice, we must not act without responsibility.[6]

The meaning of life, to Faustian man, is will and victory. The meaning of life, to the ecologically wise, is *life itself,* over which "victory" is a meaningless concept. Wisdom takes satisfaction in understanding the limits of power, how to avoid hurtful exertion within a naturally balanced whole. Knowledge is very strong, and is in many places around us transforming the world for the worse. Only character can remake and control this knowledge; only the wisdom of the centered, balanced person, who knows that life and mind are part of the earth itself.

EPILOGUE

A Personal and Historical View

If a man's imagination were not so weak, so easily tired, if his capacity for wonder not so limited, he would abandon forever such fantasies of the supernal. He would learn to perceive in water, leaves, and silence more than sufficient of the absolute and marvelous, more than enough to console him for the loss of the ancient dreams.

Has joy any survival value in the operations of evolution? I suspect that it does.

Edward Abbey, *Desert Solitaire*

A few days ago I stood in a high granite basin dotted with whitebark pines and shining with snowmelt trickles and a rushing stream. At that elevation in the Sierra Nevada range of California, the clear, thin air gives a sharpness and snap to things; outlines were clear, shadows dense, light dazzling.

I looked, and felt strangely at peace. Danger and the prospect of harm were not absent. Since I was alone on that trip and two days' walking from roads, a foolish move or a bad fall could have put me in serious trouble. Nature would not take care of me. The rules, cool and predictable, were set, and would operate as impartially upon me as upon an animal or a rock.

Nonetheless, I did belong. I felt no sense of alienation, no ache of distance between me and that lovely world of natural things. My belonging was not a sense of privilege, not a special destiny or ownership, but a simple awareness: my life had sprung from the very same earth, the identical processes, that I saw before me. Carbon, nitrogen, hydrogen, and oxygen; the history of land, water, weather, and species—all were as real and definitive for me as for the rosy little succulent *Sedum* colonizing the rock-cracks at my feet, the Oregon juncoes searching the nearby stand of pines, or this high, glacial basin itself. We were, all of us, products of a fine long history that had made thousands of beings to live in thousands of interlocked niches. We were, most of us, more alike than different. And we certainly all belonged here on planet earth. I felt glad to be the one blessed with consciousness to reflect on this huge family, this lovely intricacy. The scene that I looked on fed my mind and emotions even as it poured out air for me to breathe and water for me to drink. It made me glad—though, at that moment, I was also hungry and rather cold in wind off the pass.

○ ○ ○

Such a sense of peace has not been the typical experience, either for me or for my fellow Westerners. I'd like to tell a little about both, to show what a long road we've all come, that reaches this surprising and satisfying outlook.

When the poet Petrarch climbed Mont Ventoux in the early spring of 1336, he was doing an odd thing. Nature-love was an underdeveloped emotion in his day—almost nonexistent, by contemporary standards. Mountaineering had no standing as either amusement or sport. As Petrarch tells it, he just had an urge, a curiosity—and off he went to climb the highest mountain in his part of Provence.

In the telling, though, the poet does something even stranger to modern minds. Petrarch quickly turns his ascent into a symbol for something else. Rather than recount what he discovered of the mountain—its paths, its dangers, its eventual summit-views—he transforms the tale into a moral journey, an allegory. "I leaped in my winged thoughts," he says, "from things corporeal to things incorporeal." The climb towards the top of Ventoux becomes the Christian's struggling journey toward "the blessed life," replete with goings-astray, recoveries of resolve, narrow paths, and moral conclusions ("Wanting is not enough; long, and you attain it"). The mountain itself is nothing to Petrarch. What is real is the inward human drama. He belittles his urge to explore the summit, remembering at last that "nothing is admirable besides the mind, compared to its greatness nothing is great."[1] He gives himself over to moral reflection, and disdains his worldly entanglement with Mont Ventoux.

Though strange to modern tastes, Petrarch's separation of his experience into two distinct, unrelated components—the spiritual/human, and the physical/natural—is not really that unfamiliar. Fixation on the solitary human spirit, floating free from the world, is a basic attitude of our common Western culture. Nature is "out there," inert and unimportant except as symbol, as means, perhaps as mere stage upon which the really important human activity is played out. For Petrarch that activity was spiritual; for later generations (including ours) that activity might be economic, scientific, social, or even artistic. But the root attitude is identical: the world of nature is divided from the human world. It is a dead thing without meaning or value, except as an adjunct to the human.

When I was a young man, I resembled Petrarch (at least in attitude) more closely than I did anyone more contemporary. For one thing, I was acutely God-conscious. God watched me and I wished to please Him. The moral intensity of being continually under God's eye rendered everything else unimportant—including this natural stage, the earth, on which events unfold.

Though I was deeply interested in nature, its use to me was symbolic. I wrote poems about it—except that they were always about me. They turned out to tell about my pain, my desires, my struggles. Nature was my canvas, and I slopped it with bucketsful of purple paint, trying to express and understand my inward drama. Literature is full of this technique, which the Victorian art critic John Ruskin formally named the "pathetic fallacy." It is quite easy, as long as one neither knows nor cares much about the nature so eagerly used. The technique relies upon separation, what our times would ultimately call alienation: the human and the natural are distinct; one uses the latter as a rather arbitrary vehicle for the former. (No one seriously thinks that the morning dew weeps for Shelley's Adonais: it is just an elegant way to explore the mind of the mourner.) Since everything of interest or value is assumed to be located on the human side of the dichotomy, fancies and emotions may be freely projected onto the dead backdrop of nature.

This attitude is so rooted and characteristic in the West that we can step forward into almost any century and find examples of it. Tennyson could and did use this essential separation between the human being and the natural world. Even when he thought to really talk about nature itself, the deepest he could go was to admit the unbridgeable gap which separated it from him. Here is his awful poem "Flower in the Crannied Wall":

> Flower in the crannied wall
> I pluck you out of the crannies,
> I hold you here, root and all, in my hand,
> Little flower—but *if* I could understand
> What you are, root and all, and all in all,
> I should know what God and man is.

To understand "root and all, and all in all" is a laudable, even an ecological goal—except that the point of the poem is precisely that Tennyson *cannot* understand, cannot at all. The sentiment is bogus because, though the poet insists that "meaning" is all there in his hand, it isn't. Petrarch could read God out of the world of nature quite readily, by symbolic means. But modern times have lost the knack. For Tennyson, the experience is wistful, poignant, and (like all sentimentality) utterly without issue. He knows that no epiphany will speak. Behind the facade of romantic emotion Tennyson is a cool anatomist, detached and quizzical, running his fingers through the roots and life-hairs of the now-dead plant whose living meaning he pretends to seek. The plant is just a thing to him, as all the universe must be to the analytic and scientific eye: a dead thing, quite separate from the living viewer.

This is not just Tennyson's problem. It pervades not only Victorian nature-literature, but *most* of our literature and virtually all of our science: and in short has been the established fact of our culture (with rare exceptions) for centuries. Its most famous expression came from René Descartes, who declared that only the mind is really free and alive; all else—stones, animals, plants, even the human body—is merely mechanical. The Westerner habitually thinks only egocentrically of nature. It—nature—doesn't matter much, except as a means. Nature can be used for monetary profit, for science, for art, for literature, even for religion, but is never valued for itself.

The principal drawback of this belief has been that some feel, or felt, a little exposed by this stripping away of nature from the human. There it was, the naked soul, tremblingly alive upon an alien, lifeless shore. Nature was an It; only we, our minds, were a Thou.

For Petrarch (as for me, in earlier days) this nakedness was tolerable because the soul had God for company. The moral struggle clothed the poor thing with spiritual dignity. But for many of Tennyson's generation, this solace had worn distressingly thin; God was, they feared, rather a long way off. Science had put a lot of distance between the world and the Creator (if—O night thoughts!—if there was one). He was no longer present in every leaf and thunder-roll and heartbeat. These were understood as mechanical processes. Hence the note of near shrillness and repressed panic one detects in so much nature writing of the era, even from those as profoundly dedicated to nature as, for instance, John Muir.

The hard fact, visible on almost every page of Muir's works, is that he was a modern, a post-Romantic, in his basic assumptions about humanity, nature, and truth. His attitude toward the mountains, though filled with enthusiasm, was always that of an outsider, seeking to belong, longing for the clue that would gain him entrance.

> Perched like a fly on this Yosemite dome, I gaze and sketch and bask, oftentimes settling down into dumb admiration without definite hope of ever learning much, yet with the longing, unresting effort that lies at the door of hope, humbly prostrate before the vast display of God's power, and eager to offer self-denial and renunciation with eternal toil to learn any lesson in the divine manuscript.[2]

This passage (like many others in this book) aches with conviction of nature's primal importance. But the undoubted factual and spiritual significance of nature lay entirely *out there* for Muir. Unresting effort, self-denial, and toil are called for to wrest the secrets from the near yet remote majesty of nature. Absent is the confidence of a Thoreau that the inward journey would yield the meaning of the outward path. Absent is the informed imagination of a Wordsworth, able to penetrate the surface of nature. Muir is a thorough modern: the human subject is utterly separated from the natural object. In his pious mood, Muir adopts a more or less evangelical tone of pleading to a remote (though real) divinity. And in his scientific mood, he exposes a raging thirst for factual knowledge. In both cases, the familiar duality of Descartes between object and subject, between fact and value, has set up an invisible barrier that fatally divides Muir from his mountains. While Thoreau could find contentment and a deep, wild peace beside a tiny pond in a mere scrap of woods, not all the mountains of North America could satisfy Muir's hunger. For Muir, the real world was truly physical, and painfully remote from the spirit.

This kind of separation was a new and painful experience for those first moderns, the Victorians. One *wanted* to believe that Someone was watching the soul upon its earthly stage. One hoped that a He was behind the It, the beautiful dead It, of nature. One hoped.

The next stage in the Western relationship to nature is not hard to guess. In the face of an alien world, desperate hope eroded into plain desperation. In the twentieth century, the cool alien place that (it seemed) was our physical setting became either a hostile or a meaningless theater for the soul to play out its existential drama. God had withdrawn or died, and we were left on this damned rock with nothing to warm us but sheer courage in the face of immensity, absurdity, and death. Not a pretty picture.

As God withdrew from my imaginative life, I too wrestled with this existential loneliness. How bleak it seemed! I visited the mountains, in fact trampled and climbed and raced to exhaustion over them, trying to find a refining fire with which to burn courage out of the dross of myself. I liked high mountains best, for their very bleakness; they were perfect representations, I thought, of the struggle, the barrenness, the hostility confronting human desires. Of course, I planned, somehow, to prevail. Just like the existentialist writers whom, at first, I had not even read, I found mostly despair and anger; but, occasionally just within reach, a kind of heroism, too.

It was not enough, decidedly not enough to live on. Camus' desperate character, the "Stranger," whose story I first read in a sun-drenched meadow on the Tuolumne one summer day, tried to find it enough, but failed to convince me. At the end of the story, awaiting execution, the Stranger tells us:

> It was as if that great rush of anger had washed me clean, emptied me of hope, and, gazing up at the dark sky spangled with its signs and stars, for the first time, the first, I laid my heart open to the benign indifference of the universe. . . . all that remained to hope was that on the day of my execution there should be a huge crowd of spectators and that they should greet me with howls of execration.[3]

Was this heroism? Resignation? Peace? To me it looked like more Byronic posturing, more of Captain Ahab making grand suicidal gestures. The inward strength was a fine thing. But it led nowhere. I could not deceive myself that this was health and life. It seemed more like the very adolescent vainglory I was striving to escape.

Nevertheless, I was stuck. My puny personal drama had no audience. The stage of nature had no inherent meaning or relation to me. I longed to act and to be connected. I preferred the mountains over society, however, because there at least these facts were cold and clear, not buttered over with social pretense. I don't know how others have escaped this predicament of disconnection and alienation. I escaped it by a near-catastrophe.

○ ○ ○

The turning point came in a way I still do not entirely understand. I was leading a group of ten on a climb of Mount Jordan, a peak of 13,344 feet elevation on the Western Divide in the Sierra

Nevada. Normally this does not have to be a hazardous undertaking, though it is certainly taxing. On this day, however, a very black local cloud system was waiting in the next valley over from us. We toiled our way up the mountain under blue skies, but on reaching the summit saw the unexpected threat almost overhead already. I directed my climbers—most of them boys around sixteen—to skip lunch and abandon the peak. We ran down the backside, which was loose rock and gravel, and dashed for an easy saddle about a half-mile down the ridgeline, through which we could gain the other side of the mountain and work our way down to camp and safety.

Lightning can be deadly to those caught on rocky summits. The interval between lightning flashes and thunderclaps grew shorter and shorter as we approached the saddle. Snow and hail whitened the ridge. It seemed to take very long to get all the boys safely over the long sandy incline, in twos to minimize danger. Pair by pair they labored in the thin air, their legs churning, almost dreamlike to watch except for the sharpness of the cold and wind. At last I followed them over the little pass, and began to downclimb toward the steep, narrow channel in which the group shivered and waited. The lightning struck repeatedly around us. Then the solid granite on which I scrambled suddenly, joltingly, carried the current from a near strike right into my body through my hands and feet. I yelled out inaudibly in the wind, more in surprise than pain; the charge was diffused enough not to do any real harm.

We waited in the chute, wet and freezing from wind-driven snow, hail, and rain. We sat on our nonconducting helmets and packs. These were my climbers, my responsibility! Chance could strike any of us dead, any moment. Tears and anger and fear swept over me. "A fish fry," I thought. "A damned fish fry."

We were not unlucky. The storm passed, and soon we were descending, negotiating the rockfall hazards created by a large party in a small chute. Hours later, when we emerged onto the mountain's flank, the squall was long past. We thawed our fingers and bodies in the last bit of sun, and returned through twilight to camp.

I spent the next day alone. My climbers left; I was to await another group arriving in a day or two. I moved about my camp, which overlooked a fine lake with a snowy pass rising at the far end. Little patches of heather greened the spot, little outbreaks of flowers, a few trees. I sat long and stared.

I thought about being near death, about the mountains and their lack of concern for my well-being. I felt how silly my anger had been. Whom did I expect to rescue me? What was the sense of being frustrated with reality? It was precisely the realness of mountains that had drawn me up there. They were simply themselves. No pretenses; no exceptions. Why shouldn't they have lightning storms? A curious calm gradually settled on me—it felt, in fact, like the warm sunlight penetrating my whole body. I saw the lake, the bright faraway snow cornice, the near cliffs of Jordan, the trees and delicate vegetation. All of it seemed right, simply right. It was not hostile. It did not represent anything. Nor was it really "indifferent."

For I was not disconnected. I was subject to all the same conditions and accidents as everything I saw. I didn't feel like the Stranger, whose aim was to assert his mind and will despite the jailhouse of the cool unlistening cosmos. The fact of the world was no prison to me; it was just what I was looking for. I could assert myself all I wanted, either successfully or not. It would not change my creatureliness: my death could take place here in perfect order and harmony. My life could, also.

I was connected. I can scarcely convey that a deliverance this news was to me. I was not some alien being wandering a dead world, but a living part of the living earth. I belonged here.

○ ○ ○

It was a long path from that illumination to the place I stood a few days ago. There was, and still is, much to try to understand. But I had crossed over into a wholly new way of thinking that never deserted me. I began to learn as much as I could about the mountains. I carried pounds of books on my back as I trudged, alone or in company with others, through Sierran valleys and passes, always watching to learn how the world worked, always poring over names and species and descriptions, always looking to see where life was, how rocks lay, how the weather came up. I wanted to know, because I understood that the life of the planet was my life, and that the two could not sensibly be separated.

This breakthrough carries a modern person into new territory, into an awareness of unity with the physical world that has not been common for Westerners since before Petrarch climbed, then ignored, Mont Ventoux. For me it brought overwhelming liberation, to stop seeing the world as a moral stage separate from myself, and to stop projecting my theologies and emotions onto it. I was free at last: free to live entirely in the world, as part of it; free to accept it for itself and for no other reason; free to value What Is, and to value myself as part of What Is.

The ecological mind leaves behind the struggle to interpret nature in human terms. Instead, it interprets the human in natural terms. It finds delight, comfort, stimulus, and awe in the real complexity and order of nature. It ponders, finally, the remarkable fact that the human mind is part of that complexity, part of that order.

We are, at last, home.

Notes

INTRODUCTION

A World To Live In

1. Mircea Eliade, *The Sacred and the Profane,* trans. Willard R. Trask (NY: Harcourt, Brace, Jovanovich, 1957).

I

Earthrise and the Readiness for Ecology

1. Frank Borman, quoted in Citizens Advisory Committee on Environmental Quality, *Annual Report to the President and to the Council on Environmental Quality for the Year Ending May 1972* (Washington, DC: Council on Environmental Quality, 1972), p. 42.
2. Council on Environmental Quality, *Environmental Quality: The First Annual Report of the Council on Environmental Quality* (Washington, DC, 1970), p. vii.
3. Quoted in John C. Greene, *Science, Ideology, and Worldview* (Berkeley and Los Angeles: University of California Press, 1981), p. 13.
4. Robert Langbaum, *The Poetry of Experience* (NY: Norton, 1963), p. 11.
5. For a discussion of Social Darwinism's struggle to bring meaning out of the harshness of nature, see my essay "Social Darwinism and Natural Theodicy," *Zygon: Journal of Religion and Science* 23 (1988).
6. Herbert Spencer, *Social Statics* (London: John Chapman, 1851), p. 322.
7. Charles Darwin, *On the Origin of Species* (London: 1859; facsim. rpt. Cambridge, MA: Harvard University Press, 1964), p. 490.
8. Andrew Carnegie, *Autobiography of Andrew Carnegie* (Boston and New York: Houghton Mifflin, 1920), p. 339.
9. Frederic W. H. Myers, *Science and a Future Life With Other Essays* (London: Macmillan, 1893), p. 132. Myers is criticizing this position.
10. Thomas H. Huxley, "Evolution in Biology" (1878), in *Darwiniana: Essays* (NY: Appleton, 1908), p. 206.

11. Thomas H. Huxley, "On the Physical Basis of Life" (1868), in *Method and Results: Essays* (NY: Appleton, 1899), p. 160. Alfred Russel Wallace, *The World of Life* (NY: Moffat, Yard, 1911), p. 398.
12. Eugene Odum, *Ecology: The Link Between the Natural and Social Sciences,* 2nd. ed. (NY: Holt, Rinehart and Winston, 1975), p. 4.
13. Paul R. Ehrlich, *The Population Bomb* (NY: Ballantine, 1968, 1971, 1978).
14. Donella H. Meadows et al., *The Limits to Growth* (NY: Signet/ New American Library, 1972, 1974), pp. ix-x.
15. Meadows et al., p. 29.
16. Barry Commoner, "Beyond the Teach-In," *Saturday Review* 53 (1970), 50. Collected in Roderick Nash, *The American Environment,* 2nd. ed. (Reading, MA: Addison-Wesley, 1976), p. 239.
17. Gary Snyder, *Earth House Hold* (NY: New Directions, 1969) pp. 114, 118.
18. Charles A. Reich, *The Greening of America* (NY: Random House, 1970).
19. Petra Kelly, quoted in the *Los Angeles Times,* Sunday Oct. 2, 1983, "2 Activists Took Different Paths to Power," Section W, p. 9.
20. Mary Daly, *Gyn/Ecology: The Metaethics of Radical Feminism* (Boston: Beacon Press, 1978), p. 9.
21. Kathleen Hendrix, "A New Global Vision Called Ecofeminism," *Los Angeles Times,* April 2, 1987, pt. V, p. 14.
22. For example, see Carolyn Merchant, *The Death of Nature: Women, Ecology, and the Scientific Revolution* (NY: Harper and Row, 1980), pp. xv-xix.
23. Discussion and review of the role of feminist ideas in the ecological movement can be found in Karen J. Warren, "Feminism and Ecology: Making Connections," *Environmental Ethics* 9 (1987), 3-20, and Michael E. Zimmerman, "Feminism, Deep Ecology, and Environmental Ethics," *Environmental Ethics* 9 (1987), 21-44.
24. See the widely reprinted "Ten Key Green Values" promulgated by the 1984 St. Paul, Minnesota gathering that largely initiated the Green movement in the U.S. (calling themselves the Committees of Correspondence). "Number 7: Feminist Values. How can we replace the cultural ethics of dominance and control with more cooperative ways of interacting?" *Synthesis* no. 27 (April 1988), p. 22.

25. U.S. Bureau of the Census, *Statistical Abstract of the U.S.* (Washington, DC: U.S. Dept. of Commerce, 1972), p. 201. Alfred Runte, *National Parks: The American Experiment* (Lincoln, NE: University of Nebraska Press, 1979), p. 173.
26. Rachel Carson, *Silent Spring* (NY: Fawcett, 1962), p. 170.
27. Leonard Silk and David Vogel, *Ethics and Profits* (NY: Simon and Schuster, 1976), pp. 53, 45, and 219.
28. James S. Bowman, "Public Opinion and the Environment," *Environment and Behavior* 9 (1977), 391.

II

Holism

1. Lecture by Loren Eiseley in the series, "The House We Live In," WCAU-TV, February 5, 1961. Quoted in Ian L. McHarg, *Design With Nature* (Garden City, NY: Doubleday, 1969, 1971) p. 43.
2. J.C. Smuts, *Holism and Evolution* (NY: Macmillan, 1926).
3. Smuts, pp. 9, 15.
4. Smuts, p. 86.
5. Smuts, pp. 86, 18-19.
6. Smuts, p. 218.
7. Smuts, p. 343.
8. Aldo Leopold, *A Sand County Almanac* (1949; NY: Ballantine, 1970), pp. 272-74.
9. Lester B. Lave and Eugene P. Seskin, "Air Pollution and Human Health," in *Economics of the Environment,* 2nd. ed., Robert Dorfman and Nancy S. Dorfman, eds. (NY: Norton, 1972, 1977), pp. 433-34.
10. M. G. Royston, "Making Pollution Prevention Pay," in *Making Prevention Pay,* Donald Huisingh and Vicki Baily, eds. (NY: Pergamon, 1982), p. 2.
11. Russell H. Susag, "Pollution Prevention Pays: The 3M Corporate Experience," in Huisingh and Baily, eds., p. 17.
12. Ernest Callenbach, *Ecotopia Emerging* (1981; NY: Bantam, 1982), p. 143.
13. Eugene Odum, *Fundamentals of Ecology,* 3rd. ed. (Philadelphia: Saunders, 1971), p. 56.
14. Review of Ivan Illich, *Energy and Equity* (NY: Harper and Row, 1974), *CoEvolution Quarterly* (Summer 1974), p. 41.

15. Frederic E. Clements, *Plant Succession* (Washington [DC]: Carnegie Institute of Washington, 1916) p. 106.
16. Clements, pp. 98-99.
17. J. E. Weaver, "North American Prairie," *American Scholar* 13 (1944), 339.
18. William Morton Wheeler, "The Ant-Colony as an Organism," *Journal of Morphology* 22 (1911), 308.
19. Lynn Margulis, *Origin of Eukaryotic Cells* (New Haven, CT: Yale University Press, 1970).
20. James E. Lovelock, *Gaia: A New Look at Life on Earth* (NY: Oxford University Press, 1979).
21. Arthur O. Lovejoy discusses the *Timaeus* in this connection, in *The Great Chain of Being* (Cambridge, MA: Harvard University Press, 1936), p. 50.
22. Gregory Bateson, *Mind and Nature: A Necessary Unity* (NY: Dutton, 1979), p. 19.
23. Quoted in Robert P. McIntosh, "Some Problems of Theoretical Ecology," in *Conceptual Issues in Ecology,* Esa Saarinen, ed. (Boston, MA: Pallas/Reidel, 1980, 1982), p. 17.
24. H.S. Jennings, quoted in William McDougall, *Modern Materialism and Emergent Evolution* (NY: D. Van Nostrand, 1929), pp. 85-86.
25. Ernst Nagel, "The Standpoint of Organismic Biology," in *Conceptual Issues in Evolutionary Biology,* Elliott Sober, ed. (Cambridge, MA: MIT Press, 1984), p. 411. This is reprinted from Nagel's *The Structure of Science* (1961).
26. See Marjorie Grene, *Approaches to a Philosophical Biology* (NY: Basic Books, 1965, 1968), v-vi and *passim.*
27. Thomas W. Schoener, "Mechanistic Approaches to Community Ecology: A New Reductionism," *American Zoologist* 26 (1986), 81.
28. Michael Begon, John L. Harper, and Colin R. Townsend, *Ecology: Individuals, Populations, and Communities* (Sunderland, MA: Sinauer, 1986), p. 669.
29. McIntosh, in Saarinen, ed., p. 4.
30. William C. Wimsatt, "Reductionistic Research Strategies and Their Biases in the Units of Selection Controversy," in Saarinen, ed., p. 157.
31. Richard Levins and Richard Lewontin, "Dialectics and Reductionism in Ecology," in Saarinen, ed., p. 108.

32. Thomas B. Mowbray, "A New and Current Synthesis of Ecology" (Review of Putman and Wratten, *Principles of Ecology*), *Ecology* 66 (1985), 636.
33. R.J. Putman and S.D. Wratten, *Principles of Ecology* (Berkeley: University of California Press, 1984), p. 331.
34. Putman and Wratten, p. 331.
35. McIntosh, in Saarinen, ed., p. 31.
36. Paraphrased in Grene, p. 17.
37. *Science* 217 (1982), 718.
38. Stephen Jay Gould, "Caring Groups and Selfish Genes," in Sober, ed., pp. 123-24. This essay originally appeared in 1977. Gould reprinted it in *The Panda's Thumb* (1980).
39. Gould, in Sober, ed., p. 124.
40. Lewis Thomas, *Lives of a Cell (*1974; NY: Penguin, 1978), p. 5.
41. Thomas, p. 14.
42. Thomas, pp. 142, 14.
43. Pierre Teilhard de Chardin, "My Fundamental Vision," in his *Toward the Future,* trans. Rene Hague (1948; NY: Harcourt Brace Jovanovich, 1973), pp. 180, 166.
44. Bill Devall and George Sessions, *Deep Ecology: Living as if Nature Mattered* (Salt Lake City, UT: Peregrine Smith, 1985), pp. 142-43.

III

Balance

1. Rachel Carson, *Silent Spring* (NY: Fawcett, 1962), p. 13.
2. Carson, p. 218.
3. Carson, pp. 57. 180.
4. Joanne Omang, "In the Tropics, Still Rolling Back the Rain Forest Primeval," *Smithsonian* 17 (1987), 62.
5. Peter H. Raven, "We're Killing Our World: Preservation of Biological Diversity" (Keynote Address, American Association for the Advancement of Science, Chicago, IL, Feb. 14, 1987), *Vital Speeches of the Day* 53 (1987), 474, 475.
6. Don Hinrichsen, "The Forest Decline Enigma," *BioScience* 37 (1987), 542-43.
7. Raven, 476.

8. Bill Devall and George Sessions, *Deep Ecology: Living as if Nature Mattered* (Salt Lake City, UT: Peregrine Smith, 1985), p. 67.
9. Aldo Leopold, *A Sand County Almanac* (1949; NY: Ballantine, 1970), p. 262.
10. Council on Environmental Quality, *Environmental Quality: The First Annual Report of the Council on Environmental Quality* (Washington, DC: 1970), p. 8.
11. Robert M. May, "The Search for Patterns in the Balance of Nature: Advances and Retreats" (MacArthur Award Lecture), *Ecology* 67 (1986), 1120.
12. R.J. Putman and S.D. Wratten, *Principles of Ecology* (Berkeley: University of California Press, 1984), p. 349.
13. Putman and Wratten, p. 348.
14. Michael Begon, John L. Harper, and Colin R. Townsend, *Ecology: Individuals, Populations, and Communities* (Sunderland, MA: Sinauer, 1986), p. 773.
15. Putman and Wratten, pp. 350-51.
16. The same contrast in my textbook examples which I observed earlier applies in the issue of ecological stability. Begon et al. simply deny that stability of a community even exists (consistent with the book's general skepticism about the existence of community-level effects). Putman and Wratten offer what appears to be a more balanced picture: they work through the many miscues of the complexity/stability history, but then observe *some* ways in which complex community structure does contribute to stability. Where Begon et al. end the chapter with an absolute declaration ("There is no such thing as the stability of a community," p. 783), Putman and Wratten end with a series of open questions ("What is the role of foodweb design—of the diversity of energy relationships? Is species diversity important?" And so on, p. 355).
17. Gary Nabhan, "Kokopelli—The Humpbacked Flute Player," *CoEvolution Quarterly* 37 (Spring 1983), 6.
18. Nabhan, 5.
19. For more about Gary Nabhan, see his *The Desert Smells like Rain* (San Francisco: North Point Press, 1982).
20. Putman and Wratten, p. 59.
21. Devall and Sessions, p. 168.

22. David Pimentel et al., "Pesticides, Insects in Foods, and Cosmetic Standards," *BioScience* 27 (1977), 182.
23. Wes Jackson, *New Roots for Agriculture* (Lincoln, NE: University of Nebraska Press, 1980), p. 25.
24. Eugene Odum, *Ecology: The Link Between the Natural and Social Sciences*, 2nd. ed. (NY: Holt, Rinehard and Winston, 1975), pp. 210-11.
25. Gary Snyder, *Turtle Island* (NY: New Directions, 1974), pp. 104, 97-101.
26. Ernest Callenbach, *Ecotopia Emerging* (1981; NY: Bantam, 1982), p. 62.
27. Herman Daly, *Steady State Economics* (San Francisco: W.H. Freeman, 1977). Dennis L. Meadows and Donella H. Meadows, eds., *Toward Global Equilibrium* (Cambridge, MA: Wright-Allen, 1973).
28. Donella H. Meadows et al., *The Limits to Growth* (NY: Signet/New American Library, 1972, 1974), p. 127.
29. Snyder, pp. 99-100, 105.
30. Quoted in Daniel G. Kozlovsky, *An Ecological and Evolutionary Ethic* (Englewood Cliffs, NJ: Prentice-Hall, 1974), p. 82.
31. Snyder, p. 105.
32. Snyder, p. 97.
33. Devall and Sessions, p. 76.
34. Quoted (from memory) from the "G.E. Carousel of Progress," formerly an attraction at Disneyland.

IV

Cooperation

1. Marston Bates, *The Nature of Natural History* (NY: Scribner's, 1950), p. 108.
2. Gertrude Himmelfarb, *Darwin and the Darwinian Revolution* (Garden City, NY: Doubleday, 1959), p. 395.
3. Others came close to this insight also: for example (according to Thomas Huxley) Wells in 1813 and Matthew in 1831. But virtually all were British. Thomas Huxley, "Evolution in Biology," (1878) in *Darwiniana: Essays* (NY: Appleton, 1915), p. 222 n.
4. See Darwin's evocation of this imaginary picture in *On the Origin of Species* (London: 1859; facsim. rpt. Cambridge, MA; Harvard University Press, 1964), p. 90.

5. Charles Darwin, *The Descent of Man and Selection in Relation to Sex,* Vol. 1 (London: John Murray, 1871), p. 153.
6. Darwin, *Origin,* p. 130.
7. R.H. MacArthur, "Population Ecology of Some Warblers of Northeastern Coniferous Forests," *Ecology* 39 (1958), 599-619. For a good summary of the subject, see David Lack, *Ecological Isolation in Birds* (Cambridge, MA: Harvard University Press, 1971).
8. Paul Colinvaux, *Introduction to Ecology* (NY: John Wiley, 1973), p. 339.
9. J. David Ligon and Sandra H. Ligon, "The Cooperative Breeding Behavior of the Green Woodhoopoe," *Scientific American* 247 (1982), 126.
10. Robert Axelrod and William D. Hamilton, "The Evolution of Cooperation," *Science* 211 (1981), 1391.
11. Michael Begon, John L. Harper, and Colin R. Townsend, *Ecology: Individuals, Populations, and Communities* (Sunderland, MA: Sinauer, 1986), p. 200.
12. Warder Clement Allee, *Cooperation Among Animals with Human Implications* (NY: Henry Schuman, 1951), p. 177.
13. A governor is an engine speed regulator comprised of two weighted arms on a rotating axis: as the engine speeds up the arms centrifugally rise, cutting down the flow of fuel or releasing excess steam, and thus slowing the engine. Used as a metaphor, it is a favorite of ecological writers. It was first used to describe biological self-regulation by Alfred Russel Wallace in his 1858 paper "On the Tendency of Varieties to Depart Indefinitely from the Original Type," collected in his *Natural Selection and Tropical Nature* (London: Macmillan, 1891; rpt. Westmead, Farnborough, England: Gregg International, 1969), p. 32.
14. Eugene Odum, *Fundamentals of Ecology,* 3rd. ed. (Philadelphia: Saunders, 1971), p. 221.
15. R.J. Putman and S.D. Wratten caution, however, that it would be an oversimplification to regard predator-prey coevolution as literally a two-sided adaptation: in most real-world cases, such balancing-out occurs only with the assistance of other systematic complexities: *Principles of Ecology* (Berkeley: University of California Press, 1984), pp. 341-43.
16. Charles Singer, *A Short History of Biology* (Oxford: Clarendon, 1931), p. 318.
17. Marston Bates, *The Nature of Natural History* (NY: Charles Scribner's Sons, 1950), p. 122.

18. Singer, pp. 322-23.
19. See for example Pierre Joseph van Beneden, *Animal Parasites and Messmates* (1876; NY: Appleton, 1885).
20. Wilhelm Boelsche, *The Triumph of Life,* trans. May Wood Simons (Chicago: Kerr, 1906, 1917), p. 151.
21. Edward Step, *Messmates* (London: Hutchinson, n.d. [1913]), p. ix.
22. Bates, p. 154.
23. Though most recently, "mutualism" has gained some favor over "symbiosis" as the most generally used term.
24. Ashley Montagu, "Foreword," in *Mutual Aid: A Factor of Evolution,* by Peter Kropotkin (Boston: Expanding Horizon Books, 1955), p. 43.
25. Peter Kropotkin, *Mutual Aid: A Factor of Evolution* (NY: Knopf, 1902, 1916). Kropotkin lists seven scientists of the 1870s and '80s as sources for his work, p. 15 n.
26. Kropotkin (1916), p. 62. These words apply to natural cooperation in general.
27. Lewis Thomas, *The Lives of a Cell* (1974; NY: Penguin, 1978), p. 124.
28. Friedrich Schremmer, "Other Interactions of Animals and Plants," in *Grzimek's Encyclopedia of Ecology,* English Edition, Bernard Grzimek, Joachim Illies, Wolfgang Klausewits, eds. (NY: Van Nostrand Reinhold, 1973), p. 126.
29. Paul Feeny, "Biochemical Coevolution between Plants and Their Insect Herbivores," in *Coevolution of Animals and Plants* (Symposium V, First International Congress of Systematic and Evolutionary Biology, Boulder, CO), Lawrence E. Gilbert and Peter H. Raven, eds. (Austin, TX: University of Texas Press, 1975), p. 4.
30. Schremmer, p. 125.
31. Begon et al., pp. 180-81.
32. David Attenborough, *Life on Earth* (Boston: Little, Brown, 1979), p. 247.
33. Odum, p. 233.
34. Warder Clement Allee et al., *Principles of Animal Ecology* (Philadelphia: Saunders, 1949), p. 3.
35. Putman and Wratten, p. 318.
36. Begon et al., pp. 108, 315.
37. Richard Dawkins, *The Selfish Gene* (NY: Oxford University Press, 1976).

38. Stephen Jay Gould, "Caring Groups and Selfish Genes," in *Conceptual Issues in Evolutionary Biology,* Elliott Sober, ed. (Cambridge, MA: MIT Press, 1984), p. 122.
39. William C. Wimsatt, "Reductionistic Research Strategies and Their Biases in the Units of Selection Controversy," in *Conceptual Issues in Ecology,* Esa Saarinen, ed. (Boston: Pallas/Reidel, 1980, 1982), pp. 156, 178. In the second reference, Wimsatt is discussing R.C. Lewontin, "The Units of Selection," *Annual Review of Ecology and Systematics* 1 (1970), 1-18.
40. Elliott Sober, "Holism, Individualism, and the Units of Selection," in Sober, ed., p. 206.
41. Michael Novak, "An Underpraised and Undervalued System," in *Moral Issues and Christian Response,* Paul T. Jersild and Dale A. Johnson, eds. (NY: Holt, Rinehart and Winston, 1983), pp. 226, 229.
42. Lawrence McCulloch et al., "The Myths of Capitalism," in Jersild and Johnson, eds., p. 224.

V

Cybernetics

1. Marjorie Grene, *Approaches to Philosophical Biology* (NY: Basic Books, 1965, 1968), p. 59.
2. Henri Bergson, *Creative Evolution,* trans. Arthur Mitchell (NY: Holt, 1911). R.G. Collingwood, *The Idea of Nature* (1945; NY: Oxford University Press, 1960), p. 138.
3. For a description of vitalists' dissatisfactions with natural selection, see Charles Coulston Gillispie, *The Edge of Objectivity* (Princeton, NJ: Princeton University Press, 1960, 1973), pp. 344-47.
4. René Descartes, *Meditations on First Philosophy,* trans. Elizabeth S. Haldane and G.R.T. Ross, Great Books Vol. 31 (Chicago: Encyclopaedia Britannica, 1952), 84.
5. See above, Chapter 1, and John C. Greene's discussion in *Science, Ideology, and World View* (Berkeley and Los Angeles: University of California Press, 1981), pp. 11-14.
6. A.G. Tansley, "The Use and Abuse of Vegetational Concepts and Terms," *Ecology* 16 (1935), 284-307.
7. Walter B. Cannon, *The Wisdom of the Body* (NY: Norton, 1932), p. 24.

8. Cannon, p. 292.
9. Cannon, p. 271.
10. Eugene Odum, *Fundamentals of Ecology,* 3rd. ed. (Philadelphia: Saunders, 1971), p. 22.
11. Ludwig von Bertalanffy, *General System Theory,* rev. ed. (NY: George Braziller, 1968), p. 12.
12. Bertalanffy, pp. 11, 5.
13. Bertalanffy, p. 13.
14. Andras Angyal, "A Logic of Systems," in *Systems Thinking: Selected Readings,* F.E. Emery, ed. (Baltimore: Penguin, 1969), pp. 26, 27. This is an excerpt from Angyal's *Foundations for a Science of Personality* (1941).
15. Jeremy Campbell, *Grammatical Man* (NY: Simon and Schuster, 1982), p. 22.
16. Campbell, p. 29.
17. Quoted in Otto Mayr, "Origins of Feedback Control," *Scientific American* 223 (1970), 111.
18. Campbell, p. 31.
19. Norbert Wiener, *Cybernetics* (NY: Technology Press/John Wiley and Sons, 1948), p. 37.
20. Bertalanffy, pp. 15-16.
21. Ludwig von Bertalanffy, "The Theory of Open Systems in Physics and Biology," in Emery, ed., pp. 77, 79. This was originally an article in *Science* 111 (1950).
22. Howard T. Odum, "Ecological Potential and Analogue Circuits for the Ecosystem," *American Scientist* 48 (1960), 1. See H.T. Odum and R.C. Pinkerton, "Time's Speed Regulator," *American Scientist* 43 (1955), 331-43.
23. Wiener, pp. 71-72.
24. Campbell, p. 18.
25. Eugene Odum, p. 149.
26. Bertalanffy, "Open Systems," p. 81.
27. Kenneth M. Sayre, *Cybernetics and the Philosophy of Mind* (Atlantic Highlands, NJ: Humanities Press, 1976), p. 108.
28. For a graphic illustration, see Garret Hardin, *Nature and Man's Fate* (NY: New American Library, 1959), p. 67.
29. Eugene Odum, p. 212.
30. Eugene Odum, p. 149.
31. S.J. Putman and S.D. Wratten, *Principles of Ecology* (Berkeley: University of California Press, 1984), pp. 350-51.

32. Eugene Odum, p. 38.
33. Ramon Margalef, *Perspectives in Ecological Theory* (Chicago: University of Chicago Press, 1968), p. 3.
34. Eugene Odum, p. 251.
35. Putman and Wratten, pp. 353-54.
36. J. E. Weaver, "North American Prairie," *American Scholar* 13 (1944), 337-38.
37. Margalef, *passim.*
38. Margalef, p. 17.
39. Margalef, pp. 2, 81.
40. Margalef, p. 29.
41. Michael Begon, John L. Harper, and Colin R. Townsend, *Ecology: Individuals, Populations, and Communities* (Sunderland, MA: Sinauer, 1986), p. 630.
42. Daniel Simberloff, "A Succession of Paradigms in Ecology: Essentialism to Materialism and Probabilism," in *Conceptual Issues in Ecology,* Esa Saarinen, ed. (Boston: Pallas/Reidel, 1980, 1982), p. 89.
43. Robert T. McIntosh, "The Background and Some Current Problems of Theoretical Ecology," in Saarinen, ed., pp. 34-35.
44. Peter J. Richerson, "Limnology in Spanish" (review of Ramon Margalef, *Limnología*), *Ecology* 65 (1984), 1700. See the discussion of paradigms in McIntosh, pp. 37-38.
45. See for instance Putman and Wratten, pp. 276-77, 322-25.
46. McIntosh, p. 38.
47. Gary Snyder, "What is Meant by 'Here,'" in *Turtle Island* (NY: New Directions, 1974), p. 114.

VI

The Natural Mind

1. Jacques Loeb, *The Mechanistic Conception of Life* (Chicago: University of Chicago Press, 1912), p. 3.
2. Jeremy Campbell, *Grammatical Man* (NY: Simon and Schuster, 1982), p. 94.
3. B.G. Goodwin, "Biology and Meaning," in *Towards a Theoretical Biology,* Vol. 4 (Chicago: Aldine/Atherton, 1972), C.H. Waddington, ed., 269.
4. Campbell, p. 108.

5. Jerry A. Fodor, "The Mind-Body Problem," *Scientific American* 244 (1981), 114.
6. Discussions of two different approaches to "artificial intelligence" are Roger C. Schank (with Peter G. Childers), *The Cognitive Computer* (Reading, MA: Addison-Wesley, 1984); and Edward A. Feigenbaum and Pamela McCorduck, *The Fifth Generation,* rev. ed. (NY: Signet, 1984).
7. Fodor, 118.
8. The posthumous *Angels Fear: Towards an Epistemology of the Sacred* continues the themes explored in Bateson's last books. It is co-authored by daughter Mary Catherine Bateson (NY: Macmillan, 1987).
9. Gregory Bateson, "Toward a Theory of Schizophrenia," in *Steps to an Ecology of Mind* (NY: Ballantine, 1972), pp. 201-27. This paper was originally published in 1956.
10. Gregory Bateson, *Mind and Nature: A Necessary Unity* (NY: Dutton, 1979), p. 20.
11. Douglas R. Hofstadter, *Goedel, Escher, Bach* (NY: Random/Vintage, 1980), pp. 684-719; see especially p. 709 where he calls strange loops "the crux of consciousness."
12. Bateson, *Mind,* p. 92.
13. Bateson, *Mind,* p. 92.
14. James E. Lovelock, *Gaia* (NY: Oxford University Press, 1979).
15. James [E.] Lovelock, "Gaia: A Model for Planetary and Cellular Dynamics," in *Gaia: A Way of Knowing,* William Irwin Thompson, ed. (Great Barrington, ME: Lindisfarne Press, 1987), p. 90.
16. Lovelock, in Thompson, ed., p. 91.
17. Lovelock, *Gaia,* pp. 74-78.
18. Lovelock, in Thompson, ed., p. 92.
19. Lovelock, in Thompson, ed., p. 90.
20. Lovelock, *Gaia,* p. 146.
21. Lewis J. Perelman, *The Global Mind* (NY: Mason/Charter, 1976).
22. Bateson, *Mind,* p. 112.
23. Eugene Odum, *Fundamentals of Ecology,* 3rd. ed. (Philadelphia: Saunders, 1971), p. 283.
24. Lovelock, in Thompson, ed., p. 90.
25. Bateson, *Mind,* p. 112.
26. Kenneth M. Sayre, *Cybernetics and the Philosophy of Mind* (Atlantic Highlands, NJ: Humanities Press, 1976), pp. 65-76, 99.
27. Perelman, p. 131.

VII

Ecological Ethics

1. Aldo Leopold, "The Land Ethic," in *A Sand County Almanac* (1949; NY: Ballantine, 1970), pp. 237-64.
2. Van Rensselaer Potter, *Bioethics* (Englewood Cliffs, NJ: Prentice-Hall, 1971), p. 1.
3. Robert Disch, ed., *The Ecological Conscience: Values for Survival* (Englewood Cliffs, NJ: Prentice-Hall, 1970). Ian McHarg, "Values, Process, and Form," in Disch, ed., p. 21.
4. Donella H. Meadows, et al., *The Limits to Growth* (NY: Signet/New American Library, 1972, 1974), p. 27.
5. Wendell Berry, *The Unsettling of America* (San Francisco: Sierra Club Books, 1977), p. 90.
6. Berry, p. 222.
7. Paul B. Sears, "The Steady State: Physical Law and Moral Choice," in *The Subversive Science: Essays Toward an Ecology of Man,* Paul Shepard and Daniel McKinley, eds. (Boston: Houghton Mifflin, 1969), p. 398.
8. Harold Sprout and Margaret Sprout, *Toward a Politics of the Planet Earth* (NY: Van Nostrand Reinhold, 1971), p. 14.
9. McHarg, in Disch, ed., p. 21.
10. Bruce Allsopp, *The Garden Earth: The Case for Ecological Morality* (NY: William Morrow, 1972), p. 72.
11. Amory Lovins and L. Hunter Lovins, "The Fragility of Domestic Energy," *Atlantic Monthly* 252 (1983), 118-26.
12. Leopold, p. 239.
13. For an interesting fictional treatment of the anti-human impulse as a response to ecological crisis, see Whitley Strieber and James Kunetka, *Nature's End* (NY: Warner, 1986).
14. Charles Birch and John B. Cobb, Jr., *The Liberation of Life: From the Cell to the Community* (NY: Cambridge, 1981), p. 165.
15. But see Paul Shepard, *The Tender Carnivore and the Sacred Game* (NY: Scribners, 1973) for a surprising view advocating a return to a hunting-based social organization. (Not surprisingly, Earth First! promotes this view, also.)
16. Barry Commoner, *The Closing Circle* (NY: Bantam, 1972), pp. 140-41.
17. Dennis Pirages, *The New Context for International Relations: Global Ecopolitics* (North Scituate, MA: Duxbury/Wadsworth, 1978), p. 258.
18. Jacques Ellul, *The Technological Society,* trans. John Wilkinson (NY: Knopf, 1973), p. 21.

19. E.F. Schumacher, *Small Is Beautiful: Economics as if People Mattered* (NY: Harper and Row, 1973), p. 147.
20. Schumacher, pp. 66-67.
21. Paul Goodman, "Can Technology Be Humane?," in Disch, ed., p. 108.
22. Bill Devall and George Sessions, *Deep Ecology: Living as if Nature Mattered* (Salt Lake City, UT: Peregrine Smith, 1985), p. 141.
23. These articles have been collected in *Space Colonies,* Stewart Brand, ed., (NY: Penguin, 1977).
24. Schumacher, pp. 174-75, 194-95.
25. Schumacher, p. 173.
26. Schumacher, pp. 153-54, emphases in original.
27. Melvin A. Benarde, *Our Precarious Habitat,* rev. ed. (NY: Norton, 1973).
28. Carolyn Merchant, *The Death of Nature: Women, Ecology, and the Scientific Revolution* (NY: Harper and Row, 1980), p. 4.
29. Birch and Cobb, pp. 170-74.
30. Alan W. Watts, *The Joyous Cosmology* (NY: Random/ Pantheon, 1962), p. 58.
31. Alexander Pope, *An Essay on Man,* Epistle 1, lines 289-94.
32. Aldous Huxley explores such a proposition in his ecological utopia, *Island* (NY: Harper and Row, 1962).
33. Gregory Bateson, *Mind and Nature: A Necessary Unity* (NY: Dutton, 1979), pp. 17, 19.
34. Lawrence J. Henderson, *The Fitness of the Environment* (NY: Macmillan, 1913), p. 37.
35. Henderson, p. 273.

VIII

Collapse and Continuity

1. Daniel Yergin, "The Real Meaning of the Energy Crunch," New York *Times Magazine,* June 4, 1978. Reprinted in *Prospects for Energy in America,* The Reference Shelf, Vol. 52, No. 3, Eric F. Oatman, ed. (NY: Wilson, 1980), p. 13.
2. "Phase 1 of Carter's Energy Plan: Decontrol," *Time* (April 16, 1979), 66-68. Reprinted in Oatman, ed., p. 44.
3. James Nathan Miller, "Who Has the Facts," *Readers Digest* (January 1980), 90-92. Reprinted in Oatman, ed., pp. 29-32.

4. Committee on Nuclear and Alternative Energy Systems (CONAES), *Energy in Transition 1985-2010* (San Francisco: W.H. Freeman, 1980), p. 1.
5. William Nesbit, ed., *World Energy: Will There Be Enough in 2020?*, Decisionmakers Bookshelf, Vol. 6 (NY: Edison Electric Institute, 1979), pp. 3, 22. This book summarizes the findings of *World Energy: Looking Ahead to 2020* (NY: IPC Science and Technology Press, 1978). Both were sponsored by the Conservation Commission of the World Energy Conference.
6. Herman Kahn, William Brown, and Leon Martel, *The Next 200 Years: A Scenario for America and the World* (NY: Morrow, 1976), p. 63.
7. Yergin, p. 13.
8. Kahn et al., p. 166.
9. Kahn et al., pp. 173, 174.
10. Kahn et al., p. 151.
11. "Energy Crisis in the Campaign," *Science* 210 (Oct. 10, 1980), 164.
12. Quoted by Kelly O'Banion, "Reagan's Energy Policy: A Time for Conservation," *The Nation* (January 3-10, 1981), 19.
13. "Worse Than Watt," *The New Republic* 197 (July 6, 1987), 7-8. Carl Pope, "Undamming Hetch Hetchy," *Sierra* 72 (November/December 1987), 34-35.
14. "Stewardship Meets the Bottom Line," *Wilderness* 51 (Summer 1988), 3.
15. Riley E. Dunlap, "Polls, Pollution, and Politics Revisited: Public Opinion on the Environment in the Reagan Era," *Environment* 29 (July/August 1987), 11.
16. Dunlap, 11.
17. "Sierra Club Financial Report," *Sierra* 73 (March/April 1988), 77.
18. "The Growth in Environmental Organization Memberships," *Environment* 29 (July/August 1987), 11.
19. Donella H. Meadows et al., *The Limits to Growth* (NY: Signet/New American Library, 1972, 1974), p. 29.
20. CONAES, pp. 67, 107-112.
21. As reported in William Glasgall, "Why OPEC Is Likely To Swallow Another Price Cut," *Business Week* (Dec. 12, 1983), 89.

IX

Orthodoxy and Schism

1. Edward Abbey, "Forward," in *Ecodefense: A Field Guide to Monkey-wrenching,* Dave Foreman and Bill Haywood, eds. (Tucson, AZ: Ned Ludd Books, 1987), pp. 7-8.
2. Jamie Malanowski, "Earth First!: Monkey-Wrenching Around," *The Nation* 244 (May 2, 1987), 569.
3. Malanowski, 570.
4. Daniel Conner, "Is AIDS the Answer to an Environmentalist's Prayer?," *Earth First!,* December 22, 1987, p. 15.
5. Murray Bookchin, Letter, *Utne Reader* No. 25 (January/February 1988), 7-8.
6. Bill Devall and George Sessions, *Deep Ecology: Living as if Nature Mattered* (Salt Lake City, UT: Peregrine Smith, 1985), p. 52.
7. Devall and Sessions, pp. 65, 85.
8. Devall and Sessions, pp. 74-75.
9. Devall and Sessions, pp. 66, 76.
10. Devall and Sessions, pp. 163, 180.
11. Gregory Bateson and Mary Catherine Bateson, *Angels Fear: Towards an Epistemology of the Sacred* (NY: Macmillan, 1987), p. 11.
12. Devall and Sessions, pp. 121-22.
13. Devall and Sessions, p. 89. This is a paraphrase from unpublished work of Morris Berman. Cf. also p. 151.
14. It is puzzling that *Deep Ecology* mentions this side of cybernetics in an appendix (pp. 236-37), but otherwise consistently ignores it.
15. Devall and Sessions, pp. 139, 125.
16. E. F. Schumacher, *Small Is Beautiful: Economics as if People Mattered* (NY: Harper and Row, 1973), p. 14.
17. Quotes from *CoEvolution Quarterly* are in *News that Stayed News: Ten Years of "CoEvolution Quarterly,"* Art Kleiner and Stewart Brand, eds. (San Francisco: North Point Press, 1986), pp. 333-34, 337.
18. Devall and Sessions, p. 151.
19. Jim Cheney, "Eco-Feminism and Deep Ecology," *Environmental Ethics* 9 (1987), 115-145.
20. Cheney, 124.
21. Cheney, 124, 133n., 130.

22. Cheney, 121.
23. Cheney, 127.
24. Cheney, 132.
25. Cheney, 133, 119.
26. Cheney does glancingly acknowledge the possibility that deep ecology's metaphoric language ought not be taken too literally, mainly in a footnote in which he paraphrases Warwick Fox: "We misunderstand the intentions of deep ecologists if we insist too strongly on holding them to account on the basis of the vocabulary they actually employ" (131n.). He also looks at the possibility that deep ecology may present an ambiguous choice between metaphors: that "We may be elements of a biospherical organism *and* members of a moral community" (132, emphasis in original). In both instances, however, Cheney dismisses the point because of these authors' later reliance on "fusion" rather than "web" language.

X

The Last Dethronement

1. Alan W. Watts, *Nature, Man and Woman* (1958; NY: Random/Vintage, 1970), p. 9.
2. Paul Shepard, "Introduction: Ecology and Man—A Viewpoint," in *The Subversive Science: Essays Toward an Ecology of Man,* Paul Shepard and Daniel McKinley, eds. (Boston: Houghton Mifflin, 1969), pp. 2-3.
3. Alan W. Watts, "The World is Your Body," in *The Ecological Conscience: Values for Survival,* Robert Disch, ed. (Englewood Cliffs, NJ: Prentice-Hall, 1970), pp. 187-88. This is an excerpt from Watts' *The Book: On the Taboo Against Knowing Who You Are* (NY: Pantheon, 1966).
4. Lewis J. Perelman, *The Global Mind* (NY: Mason/Charter, 1976), p. 130.
5. E.F. Schumacher, *Small Is Beautiful: Economics as if People Mattered* (NY: Harper and Row, 1973), pp. 33-37.
6. Ursula K. Le Guin, *The Farthest Shore* (NY: Bantam, 1972), pp. 66-67.

EPILOGUE

1. Petrarch, "The Ascent of Mont Ventoux," in *The Renaissance Philosophy of Man,* Ernst Cassirer et al., eds. (Chicago: University of Chicago Press, 1948), pp. 39, 40, 44. Originally a letter to Francesco Dionigi de'Roberti of Borgo San Sepolcro, April 26, 1336.
2. John Muir, *My First Summer in the Sierra* (1911; Boston: Houghton Mifflin, 1979), p. 132. These are Muir's journals of 1869, first published in 1911.
3. Albert Camus, *The Stranger,* trans. Stuart Gilbert (1942; NY: Random/Vintage, 1946), p. 154.

Index